ADVANCING CONCEPTS AND MODELS FOR MEASURING INNOVATION

Proceedings of a Workshop

Christopher Mackie, *Rapporteur*

Committee on National Statistics

Division of Behavioral and Social Sciences and Education

The National Academies of
SCIENCES • ENGINEERING • MEDICINE

THE NATIONAL ACADEMIES PRESS
Washington, DC
www.nap.edu

THE NATIONAL ACADEMIES PRESS 500 Fifth Street, NW Washington, DC 20001

This activity was supported by the National Center for Science and Engineering Statistics of the National Science Foundation through a grant to the National Academy of Sciences. This grant (No. SES-1024012) provides support for the work of the Committee on National Statistics from a consortium of federal agencies. Any opinions, findings, conclusions, or recommendations expressed in this publication do not necessarily reflect the views of any organization or agency that provided support for the project.

International Standard Book Number-13: 978-0-309-44951-9
International Standard Book Number-10: 0-309-44951-0
Digital Object Identifier: 10.17226/23640

Additional copies of this report are available for sale from the National Academies Press, 500 Fifth Street, NW, Keck 360, Washington, DC 20001; (800) 624-6242 or (202) 334-3313; http://www.nap.edu.

Printed in the United States of America

Suggested citation: National Academies of Sciences, Engineering, and Medicine. (2017). *Advancing Concepts and Models for Measuring Innovation: Proceedings of a Workshop*. Washington, DC: The National Academies Press. doi: 10.17226/23640.

The National Academies of

SCIENCES · ENGINEERING · MEDICINE

The **National Academy of Sciences** was established in 1863 by an Act of Congress, signed by President Lincoln, as a private, nongovernmental institution to advise the nation on issues related to science and technology. Members are elected by their peers for outstanding contributions to research. Dr. Marcia McNutt is president.

The **National Academy of Engineering** was established in 1964 under the charter of the National Academy of Sciences to bring the practices of engineering to advising the nation. Members are elected by their peers for extraordinary contributions to engineering. Dr. C. D. Mote, Jr., is president.

The **National Academy of Medicine** (formerly the Institute of Medicine) was established in 1970 under the charter of the National Academy of Sciences to advise the nation on medical and health issues. Members are elected by their peers for distinguished contributions to medicine and health. Dr. Victor J. Dzau is president.

The three Academies work together as the **National Academies of Sciences, Engineering, and Medicine** to provide independent, objective analysis and advice to the nation and conduct other activities to solve complex problems and inform public policy decisions. The National Academies also encourage education and research, recognize outstanding contributions to knowledge, and increase public understanding in matters of science, engineering, and medicine.

Learn more about the National Academies of Sciences, Engineering, and Medicine at **www.national-academies.org**.

STEERING COMMITTEE AND STAFF FOR THE WORKSHOP ON ADVANCING CONCEPTS AND MODELS OF INNOVATIVE ACTIVITY AND STI INDICATOR SYSTEMS

SCOTT STERN (*Chair*), Sloan School of Management, Massachusetts Institute of Technology
ASHISH ARORA, Fuqua School of Business, Duke University
MARYANN FELDMAN, Department of Public Policy, University of North Carolina

CHRISTOPHER MACKIE, *Study Director*
ANTHONY MANN, *Program Coordinator*

COMMITTEE ON NATIONAL STATISTICS

Acknowledgment of Reviewers

This workshop proceedings has been reviewed in draft form by individuals chosen for their diverse perspectives and technical expertise, in accordance with procedures approved by the Report Review Committee of the National Academies of Sciences, Engineering, and Medicine. The purpose of this independent review is to provide candid and critical comments that will assist the institution in making its published proceedings as sound as possible and to ensure that the proceedings meets institutional standards for objectivity, evidence, and responsiveness to the charge. The review comments and draft manuscript remain confidential to protect the integrity of the process.

We thank the following individuals for their review of this proceedings: Carter Bloch, Department of Political Science, School of Business and Social Sciences, Danish Centre for Studies in Research and Research Policy, Aarhus University, Denmark, and Javier Miranda, Center for Administrative Records Research and Applications, U.S. Census Bureau.

Although the reviewers listed above provided many constructive comments and suggestions, they were not asked to endorse the content of the proceedings nor did they see the final draft of the report before its release. The review of this proceedings was overseen by Lawrence Brown, The Wharton School, University of Pennsylvania. Appointed by the National Academies of Sciences, Engineering, and Medicine, he was responsible for making certain that an independent examination of this proceedings was carried out in accordance with institutional procedures and that all review comments were carefully considered. Responsibility for the final content of this report rests entirely with the rapporteur and the institution.

Contents

1

Introduction

Because of the role of innovation as a driver of economic productivity and growth and as a mechanism for improving people's well-being in other ways, understanding the nature, determinants, and impacts of innovation has become increasingly important to policy makers. To be effective, investment in innovation requires this understanding, which, in turn, requires measurement of the underlying inputs and subsequent outcomes of innovation processes.

In May 2016, at the request of the National Center for Science and Engineering Statistics (NCSES) of the National Science Foundation (NSF), the Committee on National Statistics (CNSTAT) of the National Academies of Sciences, Engineering, and Medicine convened a workshop—bringing together academic researchers, private- and public-sector experts, and representatives from public policy agencies—to develop strategies for broadening and modernizing innovation information systems. As described in the statement of task (see Box 1-1), the workshop was organized by a steering committee to assist NCSES in refining and prioritizing its work on innovation metrics to maximize the relevance and utility of its data collection programs and statistical products to users. A background paper (Robbins, 2016) was also prepared to help identify topics of interest to address. The focus of the workshop was on a continuation of the communities' discussions and conceptualizing innovation—its inputs, dynamics, outputs, and consequences—in a way that reveals elements that are being measured well and those that are being measured inaccurately or not at all. Presenters and discussants were asked to take into

BOX 1-1
Project Statement of Task

The National Academies of Sciences, Engineering, and Medicine will convene a steering committee to organize a workshop on issues related to innovation and innovation measurement. The workshop will bring together social scientists and policy makers to review and provide input on the development of a core set of concepts and models of national and regional innovative activity and innovation systems. Discussion and the subsequent summary document prepared by a rapporteur will be oriented to help the National Science Foundation's National Center for Science and Engineering Statistics refine and prioritize its work in this area and to begin to prepare for OECD's September 2016 Blue Sky Science, Technology, and Innovation Indicators Conference.

account the role of innovation not only as it affects economic growth and productivity, but also as a mechanism for creating greater public good and meeting social challenges that are often nonmarket in nature. Workshop participants were also asked to consider how new kinds of data can be used to complement more traditional survey and administrative sources in the construction of innovation metrics. Throughout the discussions, one objective was to identify questions that cannot be answered now but could be with additional data that have a reasonable chance of being collected.

OPENING REMARKS

During his introductory comments, workshop chair Scott Stern (Massachusetts Institute of Technology) outlined why the meeting was convened and why the topical coverage is timely. He pointed out that the term "innovation" is meant to identify phenomena that are themselves important, but that they are often elusive. Nonetheless, in the context of the measurement of science and technology indicators, there is a growing consensus among practitioners and researchers who gather the data and report their results that innovation—both its inputs and potentially various elements of its outputs—is measurable and central to economic and social progress. This, Stern added, is a particularly important contention in an era when the United States and other countries are facing relatively low rates of productivity growth and periods of greater economic fluctuation. One purpose of the workshop, he said, was to bring together communities to talk about the linkage between measurement and policies that might help promote economic and social progress.

Stern identified the three major communities represented at the workshop: (1) government agencies and policy analysts who—in addition to developing large-scale statistical programs to measure various parts of the economy and society, including science and technology—are increasingly and intently focused on the measurement of the causes, consequences, and phenomena of innovation; (2) academic researchers who have developed conceptual frameworks and metrics that advance understanding of innovation; and (3) practitioners, the people who are doing the innovating. One consequence of the presence of three distinct communities is that people often use the word "innovation" to mean very different things, which can lead to an inability to establish standards that allow for clear and transparent communication. Innovation may be a complex phenomenon, and many statistics may be needed to capture its multiple dimensions. Yet it should be clear that, if communities are talking past each other, it becomes even more difficult for policy makers, practitioners, and researchers to take advantage of important new findings as they emerge.

Stern also noted the timeliness of the workshop. In addition to its long-range goals, NCSES is about to embark on a "medium-term" goal to influence the revision process by the OECD and Eurostat of the *Oslo Manual* (described in Chapter 2) and to provide input to the OECD Blue Sky Forum in autumn 2016. These activities are committed to advancing and modernizing innovation statistical programs and setting a measurement agenda for the coming decade. The workshop was intended to present broader lessons that may enhance the capabilities of NSF and other statistical agencies to continue on their path of developing a robust innovation measurement program.

Speaking on behalf of NCSES, John Gawalt (NSF) noted that, as part of its mission to conduct data collections related to U.S. competitiveness, the agency has a focused interest in innovation and, consequently, in collecting and reporting science, technology, and innovation (STI)-related information. He outlined the goals of the workshop and described NCSES activities that relate to innovation measurement.[1] He noted that the agency's history of collaboration with CNSTAT has allowed it to address challenges associated with its mandate to cover a number of topics, mostly

[1]The agency conducts and supports large-scale surveys on (1) the science and engineering workforce and the progress of STEM education and (2) R&D funding and performance across all sectors of the economy—business, academia, and, in the near future, on the nonprofit sector; the agency also collects analogous information on the federal and state governments. NCSES publishes information about the academic research infrastructure and about public understanding and attitudes toward science. Under the guidance of the National Science Board, the agency produces a biennial report to Congress, *Science and Engineering Indicators*, which publishes high-quality data on the science and engineering enterprise overall, where possible in an international context.

within the confines of the business sector and the nonprofit sector. Most recently, NCSES commissioned CNSTAT to produce the report *Capturing Change in Science, Technology, and Innovation: Improving Indicators to Inform Policy* (National Research Council, 2014) which provided an "assessment of the types of data, metrics, and indicators that would be particularly influential in evidentiary policy and decision making for the long term." The authoring panel of that report was also charged with "recommending indicators that would reflect the fundamental and rapid changes in the global STI system while having practical resonance with a broad base of users in the near, medium, and long terms" (p. 7). The report covered a broad spectrum of topics that fall within the purview of NCSES responsibilities, and included a chapter on the measurement of innovation. Gawalt noted that, in commissioning this workshop as a follow up to the *Capturing Change* report, NCSES sought to tap into participants' expertise to inform its data collection activities and to refine and prioritize its plan to improve the breadth of innovation measures. With input from the workshop and elsewhere, he said the agency hopes to shape its innovation measurement agenda for the future in a way that will benefit the user community.

WORKSHOP BACKGROUND

As referenced in the prospectus for the workshop,[2] the *Oslo Manual* (OECD-Eurostat, 2005, p. 46) defines innovation as ". . . the implementation of a new or significantly improved product (good or service), or process, a new marketing method, or a new organizational method in business practices, workplace organization or external relations." Inputs to innovation such as research and development (R&D), capital expenditures, and training are defined by the *Oslo Manual* as innovation activities that, in principle, are "all scientific, technological, organizational, financial, and commercial steps which actually, or are intended to, lead to the implementation of innovations. Some of these activities are themselves innovative, while others are not novel but play a role in the implementation of innovations" (p. 18).

Definitions in the *Oslo Manual* are oriented toward measurement of innovation by business enterprises at the firm level and in a way that facilitates international comparability. In examining the purpose of innovation measurement, Robbins (2016, p. 2) raised a set of questions about the appropriate scope of data collection objectives:

[2]This unpublished document was circulated to the workshop steering committee and to participants to communicate the goals and provide background information for the activity.

- What concepts identify the outputs of an innovation process, resulting from the activities of actors interacting with other inputs in a particular environment?
- What concepts explain the effect of different kinds of innovation over time, as with incremental innovation and more transformative or disruptive innovation, such as general purpose technologies?
- What concepts for innovation output explain industry and technology variations?
- What measures of innovation reflect the broader impact of innovation on economic growth and social outcomes?

As described in the workshop prospectus, researchers are seeking greater conceptual flexibility and bringing into scope how the successful exploitation of ideas affects the well-being of society more broadly, beyond the contribution to efficiency, effectiveness, or quality in the production of market goods and services. Knowledge creation has the capacity to influence nonmarket outcomes in areas such as health, environmental sustainability, and education.[3] Similarly, while advocating for measurement of the transmission mechanisms of new knowledge and its impacts on economic development, OECD (2010) also highlighted questions of how innovation affects the workplace, communities, and social habits.

Regardless of scope, many participants noted that measurement of innovation is difficult. Many kinds of innovations occur, which means that a simple aggregation of cases cannot be expected to be very useful for assessing (or predicting) impacts on economic growth or on other outcomes. Some innovations have a huge impact, while others are useful but minor. Additionally, not all innovation inputs and outputs can be easily quantified. Sources such as NCSES human resource data provide information necessary to measure "education, skills, and other dimensions of human capital that are used in the process of innovation" (Robbins, 2016). However, the process of R&D, for example, also involves a continuous flow of idea creation and should be measured as such. As highlighted in the workshop session on "measuring public-sector innovation and social progress," connecting measures of innovation to economic and social outcomes is often even more challenging than quantifying innovation or its inputs.

The of emphasis for this workshop—on conceptual bases for measuring innovation—was motivated in part by a key message that emerged

---

[3]For example, Cutler and McClellan (2001) found that innovations in the treatment of heart attacks, depression, cataracts, and other conditions have led to increased longevity and less absenteeism from the workplace.

from the OECD Blue Sky Forum (2006) a decade ago: that research on innovation is highly fragmented, which undermines its effectiveness. This observation led to calls for the development of general analytic frameworks and greater coordination of research efforts

In part, due to the multiple objectives of innovation measurement, workshop participants discussed at length the activities of individuals as well as those of institutions and organizations. Firm-based innovation has been carefully defined in the *Oslo Manual* (OECD-Eurostat, 2005) as "the interplay of institutions and the interactive processes at work in the creation of knowledge and in its diffusion and application." Less attention has been given to the role of individuals in innovative processes. In his address to the 2006 Blue Sky II conference (OECD, 2007), John Marburger called for new work in a way that implicitly recognized the importance of this comparatively neglected front: "We need models—economists would call them microeconomic models—that simulate social behaviors and that feed into macroeconomic models that we can exercise to make intelligent guesses at what we might expect the future to bring and how to prepare for it."

ORGANIZATION OF THIS REPORT

In summarizing the workshop, the rest of this report is organized around the topical issues presented and discussed:

- Assessing innovation measurement: How accurately are innovation processes and resultant societal and economic outcomes measured? (Chapter 2)
- What is the nature of innovation that takes place beyond R&D, and how well is it measured? (Chapter 3)
- What roles do individuals (and networks of individuals) play in innovation, and how well are they measured? (Chapter 4)
- How can public-sector innovation and innovation resulting in social progress be measured? (Chapter 5)
- What do regional innovation models tell us about innovation processes, and what are the data needs for improving measurement of at the subnational level? (Chapter 6)
- How best can innovation measurement agendas of the future be shaped? (Chapter 7)

Finally, key themes and possible future developments discussed in the closing sessions of the workshop are summarized in Chapter 8.

This report has been prepared by the workshop rapporteur as a factual summary of what occurred at the workshop. The planning committee's role involved planning and convening the workshop. The views contained in the report are those of individual workshop participants and do not necessarily represent the views of all workshop participants, the planning committee, or the National Academies of Science, Engineering, and Medicine.

2

Assessing Innovation Measurement

The workshop steering committee organized much of the content of the first day of the workshop to respond to the following set of overarching questions:

- Which aspects of the innovation processes are not well covered in current statistics and are increasing in importance?
- Among things that are currently not measured but could be, what are the top priorities? Where can funding for measurement be most productively directed?
- For what aspects of innovation measurement might conventional surveys not comprise the optimal data collection strategy? Where might administrative or big data become more prominent in meeting demand for new metrics?
- What is the path forward for measuring innovation (a difficult measurement process) in the case of public goods and social progress (difficult to measure, largely nonmarket outcomes)?

The steering committee believed that addressing these questions requires first assessing the adequacy of current measures of innovation (the topic of Session 2 of the workshop). The longstanding measurement paradigm in this field emphasizes collecting data on inputs and expenditures (e.g., investment in research and development [R&D]). While this emphasis is driven largely by feasibility constraints, it means that the impact of innovation on economic outputs and other outcomes is only captured in a partial way at best. It may also lead to excessive attribution of returns to R&D (if complementary activities tend to correlate with R&D). This measurement gap has led for calls to collect new kinds of data and create new metrics (see, e.g., National Research Council, 2014).

In this session, participants were asked to identify aspects of the innovation process and its impacts that are measured well and aspects that are being measured incorrectly, incompletely, or not at all. Such assessment encompasses inputs to innovation (R&D and other), outputs and outcomes of innovation (such as economic growth or a better functioning society), and the utility of metrics to stakeholders ranging from government to researchers to businesses themselves. This discussion was intended to help the National Center for Science and Engineering Statistics (NCSES) assess the adequacy of current data collection programs and what additional kinds of information should be pursued.

THE PURPOSE OF INDICATOR FRAMEWORKS AND UNDERMEASURED ASPECTS OF INNOVATION

Bronwyn Hall (University of California, Berkeley) began the session by reviewing desirable characteristics of innovation indicators in the context of how they are used for policy and other purposes; she also commented on existing data gaps. Drawing from Hall and Jaffe (2012), she pointed out that one important use of current indicators is for benchmarking. Since it is difficult to measure returns to innovation activities, organizations set goals of how to spend on a range of activities by benchmarking and carrying out performance assessments based on changes in these benchmarks. This need requires data that are comparable across space and over time. Innovation indicators are also used for public policy questions, private-sector decisions, and academic research; for the latter, microlevel information and data matched to other sources (an area for which there has been recent progress) are desirable.

Hall pointed out that, until now, the dominant data collection methodology has been surveys—for example, asking companies how much they spend on research, or asking individuals about their education and career paths. However, due to falling survey response rates and increasing costs, there is a growing sense that some data might be better collected using administrative and other passive methodologies that utilize data created as a matter of course for operational or other (nonresearch) purposes. One example is accounting data maintained by firms that are sometimes available publicly. The patent system also generates passive data, and, again, these are public sources. One advantage of passive data, Hall said, is that if they are generated during the course of other activities, they lend themselves less to gaming by respondents. She said a key issue to be addressed in using passively collected data to complement survey data is quality, which is important for web scraping and other emerging methods. As noted in Hall and Jaffe (2012), for passive methods, "it may be much harder to assess reliability, precisely because the relationship between the captured data and the universe of underlying activity is not known" (p. 8). An additional issue to be addressed by statistical agencies, and which will systematically differ between survey and nonsurvey data and between publicly and privately generated data, is accessibility—an issue considered at various points throughout the workshop.

One approach used for measuring innovative activities is the growth accounting framework, Hall explained. This framework is used for measuring R&D, which is only a subset of innovative activities in the National Income and Product Accounts (NIPAs). The difference between the growth accounting exercise done for the NIPAs and what microeconometricians do is that the former is basically an input cost approach. Normal rates of return are assumed, which may not be wholly appropriate for intangible inputs like R&D, and, from a welfare perspective, the value of unpriced output is omitted. Additionally, the aggregate accounting approach is something of a black box that obscures the functional relationships underlying the system. In contrast, she noted, researchers use the microeconometric approach to try to find out how much innovative activities contribute to the value of output (whereas the growth accounting version assumes that the value of the contribution is equal to what is spent on it). For many questions, it is the value of the output of innovative activity that is most ideal to measure.

Having more and better indicators creates opportunities to examine variables in models of innovation as well as feedback mechanisms that exist, such as the influence of ideas coming from users back to firms. The impacts of basic science discoveries that emerge from applied research are also important, but predicting the flow of information in this way is extremely difficult. Hall and Jaffe (2012) concluded that, at least for the United States, indicators of resource flows—inputs to innovation—are fairly well covered, but that flows within government, industry, academic, and household sectors are less well measured. And non-R&D inputs are typically not measured at all, at least in the United States.[1] This gap means that measures of innovation expenditures could be off by orders of magnitude if the goal is to measure all innovation. Human capital formation and knowledge output is measured fairly well, but the proxies that they provide are distant from the underlying concept of interest to researchers and policy makers. For example, counts of papers, degrees, and patents

[1]European Union member states have had programs to measure expenditure on non-R&D innovation activities.

are available, but the success or impact of them (e.g., the value of outputs linked to them) is less well measured.

Hall concluded that to address gaps in data coverage, improving the ability to link sources, to generate timely data, and to produce information on capital for financing innovation (much of which is privately held) should be top priorities for the statistical agencies.

Ben Martin (University of Sussex) identified the challenge for innovation measurement, which is to lessen the danger that it fails to keep up with the changing world and the changing nature of innovation. NCSES, other statistical offices, and the OECD play an important role in providing the empirical infrastructure for the field of science, technology, and innovation policy studies. The national and international research system—in which government actors, university researchers, and the private sector all play a part—must work together to meet this challenge.

Martin highlighted one area for which innovative activities go largely unmeasured: the knowledge economy. While encouraged by emerging opportunities created by new kinds of data, Martin pointed to a number of challenges. Increasingly, the problems of interest, such as growing competition and increasing complexity of productive processes, which put a growing premium on innovation, science, and technology, are global rather than national or regional. Innovation takes place in a vast range of sectors (not just in manufacturing, as was often implicitly assumed until the 1990s) and in different kinds of organizations (not just firms, again as was often assumed in the past). Additionally, he said, innovation is not just technological; for example, it may be process oriented.

The rapidly changing environment motivated one central question addressed by many presenters during the workshop: Are current innovation indicators adequate? Citing Hall and Jaffe (2012), Martin prefaced his answer by noting that all indicators are partial. Especially in the world of social science and policy, indicators only capture certain aspects of a phenomenon or its characteristics, and then only to a limited extent. Sometimes, he noted, caveats are forgotten, assuming, for example, that patents measure innovation, which they do not. This approximation ignores the subtle distinction between innovation and invention, as well as the fact that patents are only used in certain sectors for certain technologies and for certain types of innovations.

Martin noted, for a given indicator, conceptual clarity is needed about what aspects of innovation are being captured and what aspects are being neglected. For example, bibliometric indicators relate to only one form of scientific output: publishing. But a whole range of other outputs from research may be equally important, such as commercial applications. Likewise, citations relate to impact on academic peers rather than to quality or outcomes that affect the balance sheets of firms or the well-being of individuals.

Martin next observed that every indicator is based on assumptions, many of which are implicit and rarely subject to critical scrutiny. The validity of these assumptions varies with circumstances and over time. For these reasons, he characterized statistics and indicators as more of an art than a science. Reiterating a point made by Hall, he noted that outdated measures are often maintained due to the appeal to researchers of consistent time series. When NCSES redesigned its R&D survey in the early 1990s to send to a much larger group of firms, the change required extra work to interpret trends spanning the pre- and post-change periods. However, dogmatically sticking with an outdated approach can be counterproductive when the nature of phenomena of interest changes (e.g., the types of individuals or organizations innovating) or the relevant population changes (e.g., a shift to embrace services as well as manufacturing firms). Sometimes, Martin asserted, maintaining a long time series comes at a cost in terms of growing distance from reality.

To the extent that innovation is conceptualized, defined, and measured in a way that reflects conditions when the field of innovation studies was being established—when most innovation was assumed to be technology-based and conducted by private-sector firms in the manufacturing sector, especially in high tech—today's indicators are failing to keep up with the changing world. In the 1960s, 1970s, and 1980s, technological innovation may have been better captured by indicators such as those for R&D spending (although R&D has never been a satisfactory proxy for all innovation), numbers of qualified scientists and engineers, patents, and so on. But now a huge amount of innovative activity takes place that is not technological, not based on R&D, not reflected in patents, and not in the manufacturing sector. Citing one example, Martin observed that the innovations that most impacted people's lives in the first decade of this century were financial—mortgage-backed securities, collateralized debt obligations, credit default swaps, and the like. Neither these innovations nor their negative outcomes were captured by conventional indicators and, as a result, they were not investigated by the innovation studies community.

Martin stressed that a lot of innovative activity is largely invisible through the lens of current measures. Here, he drew an analogy with cosmology, and the fact that about 95 percent of the universe is invisible with current measuring techniques (telescopes). It takes the form of dark matter or dark energy. For this reason, he adopted the term "dark innovation," that which is invisible with current metrics. In addition, unmeasured innovation may be taking place disproportionately in places, such as China, Latin America, and Vietnam that are producing a large percentage of goods consumed worldwide. Much of this innovation is not being captured in any official statistics, which further distorts our perception of what is happening.

During open discussion, Scott Stern called for more evidence to assess whether the rate of dark innovation had been changing in recent years. If it has been increasing, that means that the measurement system has become more partial or grown more outdated. As noted above, the original measurement system for innovation, when set up, was perhaps well designed for the phenomena at that time. But since the phenomena have changed, Stern said evidence is needed regarding the extent of dark innovation, and there are currently little hard data on which to study this question.

Pointing to the challenges now being faced, Martin identified the need to conceptualize, define, and devise methods for measuring dark innovation. Some solutions may be facilitated by opportunities created in an era of big data. He warned, however, about several dangers. The first of these is characterized by the metaphor of a person looking for his keys under a lamppost. A temptation of indicator producers will be to focus on phenomenon and characteristics where the light is, which could lead to the neglect of less easily measured or indeed nonmeasurable aspects, even if they are equally or more important. This methodological tendency is also linked to the McNamara Fallacy: making the measurable important rather than attempting to make the important measurable.[2]

A second danger identified by Martin is Goodhart's Law, which states when a variable is adopted as a policy target, it rapidly loses its ability to capture the phenomenon or characteristics supposedly being measured.[3] In this context, once an indicator is adopted as part of policy, it leads to changes in behavior—specifically, game playing may take place in response to perverse incentives, thereby creating unintended consequences. As an example, Martin cited the British minister for science talking about the number of spinoffs from universities. When he started talking about it, there were 70 spinoffs a year; the next year, the number grew to 200. Clearly the universities had not become three times better at this activity: they had just become better at playing the game.

The third danger facing innovation measurement according to Martin has to do with costs and benefits. Indicator development entails setting indicators up and regularly updating them. In some cases, the costs,

[2]Named after Robert McNamara, U.S. Secretary of Defense, 1961 to 1968; first coined by Daniel Yankelovich in 1972.

[3]Named after Charles Goodhart, a former advisor to the Bank of England and emeritus professor at the London School of Economics, who first gave expression to Goodhart's Law in 1975.

which may include distortions created by gaming the system, may come to exceed the benefits. Martin sees this as possibly being the case in the application of bibliometric indicators, where individuals aiming to score in this system have incentives that may lead to higher incidence of research misconduct.

Martin concluded that in a knowledge-intensive society, innovation becomes even more important than it has been in the past, but it takes on a growing variety of forms and geographic characteristics. Conventionally accepted indicators reflect primary forms of innovation of previous decades, which has left much current innovative activity invisible or dark. Thus, new indicators are needed. But, in an era of easily available big data, the temptation to search only "under the lamppost" must be resisted. Martin urged continuing awareness of the McNamara Fallacy and of subsequent game-playing and unintended consequences, and, above all, that the benefits of indicator systems must be greater than their costs.

INTERNATIONAL INDICATOR STANDARDS—THE *OSLO MANUAL* AND THE COMMUNITY INNOVATION SURVEY

Fred Gault (United Nations University–Maastricht Economic and Social Research Institute on Innovation and Technology) spoke to a range of issues: developing new indicators to capture the changing nature of innovation; policy and other uses of innovation data and indicators; and international comparisons. His approach to the challenge posed by the workshop—to identify questions that cannot now be answered but could be with additional data that have a reasonable chance of being collected—was to compare what can be done within the existing definition of innovation in the *Oslo Manual* with what could be done if changes were made to the definition and new data sources were then developed.[4] The current definition in the *Oslo Manual* is as follows:

> An *innovation* is the implementation of a new or significantly improved product (good or service), or process, a new marketing method, or a new organisational method in business practices, workplace organisation or external relations (OECD-Eurostat, 2005, para 146).

Further:

> A common feature of an innovation is that it must have been *implemented*.

---

[4]For background material and references underlying Gault's arguments, see Gault (2016). *The Oslo Manual: Guidelines for Collecting and Interpreting Innovation Data* (3rd ed., 2005). Available: http://www.oecd.org/sti/inno/oslomanualguidelinesforcollectingandinterpretinginnovationdata3rdedition.htm [August 2016].

> A new or improved product is implemented when it is introduced on the market. New processes, marketing methods, or organizational methods are implemented when they are brought into actual use in the firm's operations (OECD-Eurostat, 2005, para 150).

The new or improved product does not have to make money to be an innovation—it just has to be put on the market. And similarly, new processes, marketing methods, and other changes are implemented when they are brought into actual use in firms' operations.

Gault echoed comments by Hall and Martin that the world in which innovation occurs (and is measured) has changed enormously over the years. About one-half of the economy consisted of manufacturing or services related to the distribution and sale of goods (wholesale and retail trade, transportation, and warehousing) in 1947 when the Bureau of Economic Analysis Gross Domestic Product (GDP) series began. Today, he said, service industry innovations in sectors, such as financial services, health care, computing services, and others dominate in a similar fashion and have transformed people's lives (though not always for the better, as the 2008 financial crisis made clear). In addition to the economy itself changing, the classification of industry for statistical purposes, as reflected in the North American Industry Classification System, has likewise changed.

The *Oslo Manual* was first published in 1992 and drew on about 20 years of experimental surveys from Canada, Germany, Italy, the Nordic countries, and the United States. The *Manual* was initially concerned with only manufacturing and technological innovation. In the second edition, published in 1997, coverage expanded to cover private-sector services, but the emphasis remained on technological product and process innovation. Published in 2005, the third edition recognized nontechnological forms of innovation, especially organizational, managerial, and marketing, for the first time. Problems identified by Gault include that types of innovation are not exclusive. Not infrequently, product innovation involves process innovation, organizational change, and marketing. An innovation can also be distributed across a range of activities, which overlap. A systems approach has been creeping gradually into the *Manual*, which is essential for understanding a complex system. The idea is to better measure linkages and feedback of different kinds of innovation. Some of these ideas will appear in the forthcoming revision to be issued in 2017.

The Community Innovation Survey (CIS)—a series of harmonized surveys based on the *Oslo Manual* and fielded by national statistical offices throughout the European Union—provides data on sources of information for innovation and on collaboration for innovation.[5] One thing that is missing, Gault pointed out, are data on sources of prototypes, such as a whitewater kayaker modifying a kayak in a way that feeds back into future design. Data could also be considered, either survey or administrative, on the size and age of firms. Sources such as business registers also provide data on firm entries and exits by size and by innovation status. Gault noted that this information can be and often is coordinated and overlaid with additional indicators based on analyses of sources of innovation, human resources, characteristics of innovators, and other factors. However, problems often exist with regard to the quality of matches across sources.

Another point raised by Gault and several others is that innovation does not always relate to R&D. Arguably, one of the most significant findings of the CIS is the extent to which firms innovate without engaging in R&D. NCSES data indicate that firms performing R&D have a substantially higher propensity to innovate; however, there are still more than 1 million "non-R&D" firms in the U.S. economy reporting innovation. This number raises a policy question that is not being addressed: how to prompt innovation—perhaps through state and municipal support—among non-R&D firms operating in different sectors. Gault referred to Martin's "dark matter" problem—if firms are not engaging in R&D, then how are they innovating?

Gault speculated about how measurement might be broadened by extending the *Oslo Manual* definition to cover innovation beyond the business sector, to the public (government) and household sectors, in a way that makes it possible to examine how innovation links together economy wide. He suggested a new definition might take the following form:

> An *innovation* is the implementation of a new or significantly changed product or process. A product is a good or a service. Process includes production or delivery, organisation, or marketing processes. A new or significantly changed product is *implemented* when it is made available to potential users. New or significantly changed processes are implemented when they are brought into actual use in the operation of the institutional unit, as part of making product available to potential users.

The main change he proposed involves generalizing the language so that "process" has three components that overlap; "improved" becomes "changed" to avoid normative statements; and "market" becomes "potential users." In the new definition, implementation occurs when the product is made available to potential users, which also applies in the public

---

[5]See http://ec.europa.eu/eurostat/web/microdata/community-innovation-survey [August 2016].

sector as well as the business enterprise sector. In addition, the process activities involve making the product available to potential users.

In discussion, Hall commented about the importance of measuring whether anything actually happened once a new or improved product is on the market (e.g., the share of sales due to products introduced during the past 3 years). This is a question that many firms are able to answer, at least approximately. Survey developers, she argued, should continue to give thought to how questions might be developed in order to generate more useful variables about the impact of innovations, rather than the yes/no information concerning whether the organization introduced an innovation. Data on market outcomes are quantitative and more informative, especially when comparing firms of different sizes, in part because they are size normalized. Martin commented that such a question brings time scales to the surface, because when a firm is asked whether it has put products on the market, the firm is able to look back over a specific reference period and report that new products that went on the market accounted for *x* percentage of sales. There are analogous metrics for the public sector—for example, for health care, perhaps decreased bed days or gains in quality-adjusted life years created as a result of a new or significantly improved product or process. Surveys have to be customized to capture the value of innovation for the public sector, which would make them very different from the CIS, which is the same regardless of sector.

A final point made by Gault is that the goal of policy makers is not only to talk about or measure innovation, but also to nurture good innovation, green innovation, and sustainable innovation. He argued that current statistics are designed to count innovations, not to ask whether they are good or bad. In order to do that, agencies have to conduct surveys after the fact, which requires a long-term strategy for measurement and a commitment to understanding the system.

Fernando Galindo-Rueda (OECD), following up Gault's discussion of the *Oslo Manual* revision, considered international standards for measuring and comparing innovation. As he and Gault outlined, the OECD-Eurostat *Oslo Manual* embodies, as the outcome of a global consensus-building exercise, a number of measurement choices reflecting sectoral scope, perspective on the unit of interest, data sources, who is performing the measurement, and who benefits from the data. Galindo-Rueda's presentation focused on the role of international standards for measuring innovation and the scope for extending them in order to facilitate the international benchmarking of U.S. innovation performance—one of the potential NCSES requirements for data on innovation.

Standardization requires above all a community of practitioners with defined responsibilities and resources who have shared interests about what to measure and how to measure it. The *Manual* reflects the perspective of national statistical organizations, although many measurement systems involve partnerships with academic researchers, who carry out data collection, and with international organizations, who themselves field surveys across the world. Finding a common denominator among stakeholders, Galindo-Rueda noted, is not trivial because quite often individual countries have idiosyncratic perspectives on why innovation is of interest and what aspects are worth measuring. In some cases, phenomena of interest are universal, and this is increasingly becoming the case as the world becomes more global. In other cases, where interests and practical constraints are location specific, it is more challenging to compare results, such as outputs of innovation, across countries.

The *Oslo Manual* provides guidelines oriented toward the measurement of innovation through statistical surveys based on self-reported measures from respondents (who are typically managers within firms). It focuses on businesses, although there has been a struggle to reach all activities (e.g., agriculture or knowledge-providing services). A number of alternative perspectives could be adopted to measure innovation, Galindo-Rueda suggested. In terms of sectoral scope, surveys of households, governments, and nonprofit organizations could all be fruitful additions.

The perspective of the *Oslo Manual* is on measuring dimensions of innovation that are descriptive of the institutional units—that is, firms—being covered. One advantage of this approach is that it creates the possibility of linking information on innovation at the level of the firm with information about firms' economic outcomes. But there are alternatives such as to focus on the actual innovations: an object-based approach, which was prominent until the early 1980s, or on innovation projects or innovation related to transactions, and perhaps inventions as a subset.

To date, statistically representative surveys have focused on eliciting information from management, but they could be reoriented toward capturing relevant insights from other perspectives such as the firm's workforce or from outsiders such as users of products and services. The methods prescribed by the *Manual* generate information about innovation as reported by a firm; but, perhaps information is also needed from users to verify whether those self-assessments are sufficiently accurate and reliable.

Regarding data sources, the *Manual* could in principle expand beyond statistically representative surveys to cover use of administrative and commercial data, and new platforms that have the potential to be used by researchers to generate interesting insights. Specialized or nonspecialized media are also potentially useful data sources. There is a history of research in the academic literature based on announcements in the media about new product launches and similar activities. One can envision new indicators based on data from the Kickstarter platform, for example, which may be used to identify what proportion of projects receive funding and how much. Some of these new data gathering and analysis activities will no doubt develop outside of the confines of the *Oslo Manual*. Some of these measurement avenues could be the object in the future of standardization, but it is important to note that not all areas are equally mature for that purpose, bearing in mind the OECD criteria for endorsing statistical guidelines. In considering new kinds of data and guideline development, OECD will no doubt demand quality assessment along the lines described by Hall and Jaffe (2012); NCSES may also play a similar role in the U.S. context for subjects of high domestic relevance where no international consensus exists. Incorporating new kinds of data is also likely to require experimentation and building up different types of consortia that involve different actors that take advantage of specialization and comparative strengths.

INNOVATION AND PRODUCTIVITY: RECENT PUZZLES

Dan Sichel (Wellesley College) discussed measurement of innovation within a macroeconomic framework, specifically what national-level and sectoral-level productivity statistics tell about innovation. A key question in this context is whether the productivity statistics currently being produced by statistical offices are credible and, in light of the answer, what might be the next steps forward.

National-level data imply that the pace of innovation has been exceptionally slow in recent years. Sichel expressed the view that this conclusion is not credible, and identified some key areas in which data inaccuracies need addressing, most notably in price measurement. While acknowledging that the productivity framework takes on a black-box characteristic, he argued that the numbers that have emerged have affected the narrative in the popular and financial press about innovation and affect the questions on which researchers focus.

If done correctly, Sichel said, the framework for measuring multifactor productivity produces (at least indirect) estimates of rates of innovation or technical change, both for the overall economy and within key sectors. In the "primal approach"[6] that begins with a production function and information about outputs and inputs, the part of growth that is not explained by changes in quantity of inputs is attributed to multifactor productivity growth that reflects productivity improvements. The approach can be used to decompose multifactor productivity growth by key sec-

[6]The production function takes the form $\Delta Y_t = \alpha\Delta K_t + (1-\alpha)\Delta L_t + \Delta \boldsymbol{mfp}_t$, where Y = output, K = capital inputs, and L = labor inputs.

TABLE 2-1 Multifactor Productivity Growth for Selected Sectors

	Average Annual Percentage Change			
	1974-1995	1996-2003	2003-2010	2010-2015
Total	0.6	2.1	0.7	0.4
High Tech	12.3	18.0	12.2	5.8
Semiconductors	28.0	47.0	27.7	6.6
Computers	15.7	13.7	8.6	8.3
Communications	1.7	0.4	0.6	3.1
Intellectual Property	2.1	2.9	2.0	1.6
Software	5.8	4.2	3.0	2.5
R&D	0.5	1.7	0.6	0.3
ELA originals	0.8	1.9	2.3	1.7
Other	0.3	1.4	0.4	0.3

NOTE: R&D = research and development; ELA = entertainment, literary, and artistic.
SOURCE: Workshop presentation by Dan Sichel, May 19, 2016; estimates derived using approach of Byrne et al. (2013) and data from Fernald (2012), updated by Sichel to 2015 with data from the Bureau of Economic Analysis, Bureau of Labor Statistics, and U.S. Census Bureau.

tors, as described by Byrne et al. (2013, 2015). Sichel noted an alternative way to measure multifactor productivity, which involves explaining price changes for an output in a particular sector, such as semiconductors. Price changes that are not explained by a weighted average of input costs also indicate multifactor productivity growth. The faster prices fall, the more rapid will be measured multifactor productivity growth.

The data that feed into these exercises are put together by the Bureau of Economic Analysis, the Bureau of Labor Statistics, and the U.S. Census Bureau. Table 2-1 presents rough first-pass estimates (from an upcoming paper) for the United States, over selected periods, of growth rates of multifactor productivity

If it is assumed that productivity figures provide reasonably good proxies for rates of innovation, then in these sectors, there is a clear pattern: a big step up in rates of productivity/innovation in the mid-1990s through the mid-2000s (emergence of the Internet, computerization, and so on) to 2.1 percent, followed by a step down, and then an even further step down since 2010. The same pattern observed for the high tech sectors is also present for intellectual property products and in other areas of the economy. Taken at face value, this means that rates of U.S. innovation, particularly in the past 5 years, have been exceptionally slow. This observed trend has colored the narrative of what is happening in the economy, as illustrated by macroeconomist Robert Gordon (Gordon, 2016). The slow pace of innovation is one of the factors Gordon cites for why future prospects for improvements in productivity and living standards are so bleak in the United States.

Sichel argued that this accounting is only credible if the data feeding into the national system for measuring productivity are accurate. For example, if the rate of price decline of semiconductors is understated, so, too, will be the rate of multifactor productivity growth since the rate of multifactor productivity growth is being inferred from rates of decline in prices of key products. Further, if the rates of multifactor productivity growth are understated, the implicitly inferred rates of innovation in these sectors will be underestimated. Solow made a statement in the late 1980s about computers, which can be paraphrased to capture the current situation: we see innovation everywhere but in the productivity statistics.

According to Sichel, many people have a sense that the productivity numbers derived from official statistics may not be up to the measurement task, and particularly that price measurement has not kept up with rapidly changing goods and services. A set of papers convincingly documents the evidence, at least within the high tech area and for some intellectual property products. For example, research on semiconductors by Byrne et al. (2013) suggests that rates of price decline are understated by as much as 35 percentage points in recent years (see Figure 2-1). Dramatic misstatement of rates of price decline for semiconductors would lead to incorrect inferences about innovation in the sector. In addition, similar research (e.g., Corrado et al., 2011) has identified measurement problems with official statistics for a range of other high tech and intellectual property products as well, suggesting that mismeasurement of multifactor productivity is broader than just in semiconductors.

Once a revised semiconductor price measure is fed into the broad system of productivity measurement, the inferred rates of innovation change sharply. Sichel noted that the blue bars in Figure 2-2 are rates of multifactor productivity growth within the semiconductor sector by period, using official price measures. These data indicate a sector that had its heyday in the mid-1990s. The red bars reflect the corrected price measures and present, in turn, a quite different picture of what has happened with multifactor productivity (and, by extension, innovation) in this sector. Specifically, innovation in this sector has continued at a rapid pace—not quite as rapidly as during the second half of the 1990s, but nonetheless quite significantly.

With fully corrected statistics, Sichel argued that researchers and policy makers might be having a different conversation and asking different questions. The conclusion that innovation in the economy is very sluggish raises one set of questions, for example, about what to do about it. In contrast, the conclusion that innovation has been quite rapid but is not appearing in broad economic statistics raises a very different set of questions. They imply two very different lines of research, with a different set

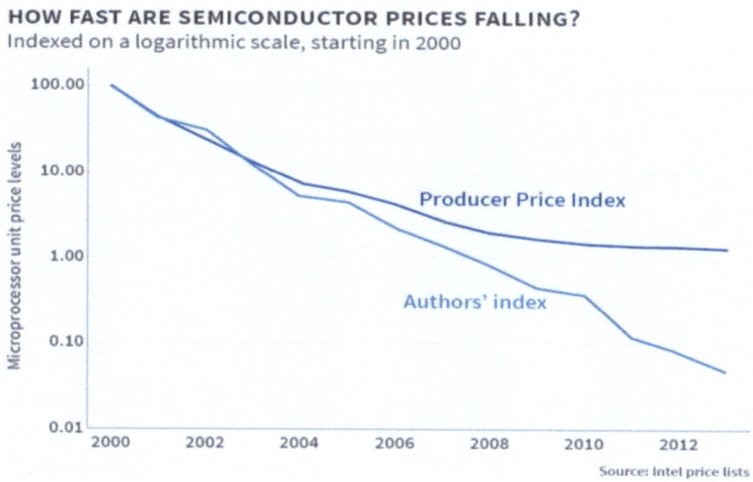

FIGURE 2-1 Price decline for semiconductors, based on two different price indexes.
SOURCE: Workshop presentation by Dan Sichel, May 29, 2016; graph from the National Bureau of Economic Research, *The NBER Digest* (July 2015, p. 5). Available: http://www.nber.org/digest/jul15/jul15.pdf [August 2016]), which was adapted from Byrne et al. (2013, Fig. 9).

of policy interventions. Even though the broad analytical framework is a bit of a black box, Sichel noted, it is important to try to make progress to improve the sorts of inferences about innovation made from that broad system.

He suggested that to improve the accuracy of multifactor productivity measurement and, in turn, inferences about innovation in key sectors, work on improving measures of prices within key sectors needs to continue and to be incorporated into the measurement programs that feed into official statistics. He also suggested an expansion of the measurement system to better capture ongoing innovation activity.

Responding to Sichel's presentation, Ashish Arora (Duke University) expressed concern about equating advances in multifactor productivity with advances in innovation. He pointed out that macroeconomic measurement is limited to revealing something about the rate at which inputs are being applied to economic production; this does not mean innovation in the conventional sense. Sichel responded that John Fernald (Fernald, 2012; Fernald and Jones, 2014) has been as careful as possible with aggregate data to account for influences such as changes in utilization that might affect multifactor productivity and that are not associated with

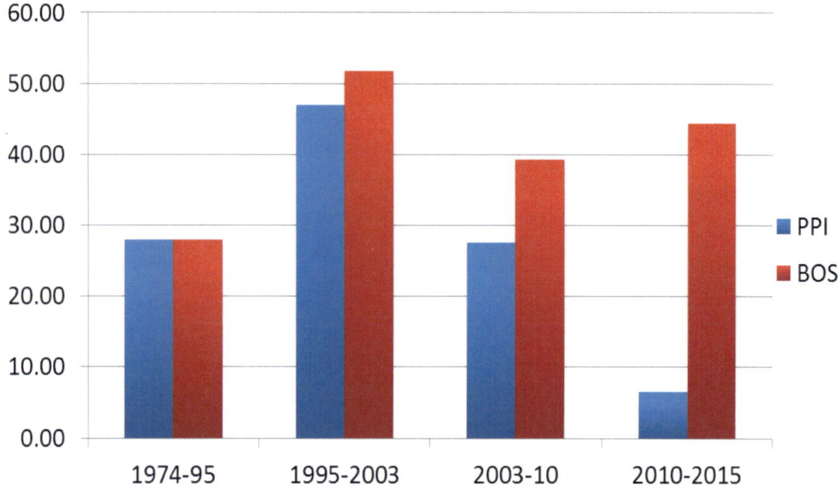

FIGURE 2-2 Two multifactor productivity measures for semiconductors (percentage change over period shown).
SOURCE: Workshop presentation by Dan Sichel, May 29, 2016; based on updating Byrne et al. (2013) using the Producer Price Index (PPI) and the Byrne et al. (2015) index (BOS) for semiconductor prices.

innovation—in other words, to construct a residual that is as close as one can get. Sichel acknowledged that productivity is not a perfect measure of innovation, but that innovation is typically viewed as one of the key factors driving it. While Sichel agreed with Arora that innovation is not the same as advances in human knowledge, he said semantic issues need to be worked out. Other factors that contribute to productivity change should also be looked at going forward by researchers. The advantage of multifactor productivity is that it offers a consistent framework within which to make comparisons across sectors and countries, and innovation is going to be one of the important factors driving those numbers.

Building on this exchange, Wesley Cohen (Duke University) pointed out that bridging the microlevel view often used by innovation researchers with the multifactor productivity view of macroeconomists reveals a need to measure additional multiple dimensions of productive processes. Technology and idea adoption, diffusion, and imitation (which can also show up in the productivity numbers) all need to be measured since productivity gains from innovation can be generated through any of these avenues. This information is necessary, Cohen said. National statistical agencies have occasionally taken on these tasks, but not necessarily in a systematic way.

3

Innovation Beyond R&D and Conventional Input Measures

Data on specific aspects of innovation—such as research and development (R&D) expenditures, patent applications, and citations—are often used as proxy levels of innovation activity within industries or countries. However, in the measurement community, it has long been understood that activities, both technical and nontechnical, including those that are not a directly measurable function of R&D and science, also play an important role in innovation (see, e.g., National Research Council, 2014).

The annual Business R&D and Innovation Survey (BRDIS)—conducted by the Census Bureau through an interagency agreement with the National Center for Science and Engineering Statistics (NCSES)—asks firms in manufacturing and service industries for information about newly introduced or improved products and processes, and about purchased R&D inflows and outflows, revenue from the sale of patents and patent licensing, and a number of activities related to intellectual property (IP). Some questions about innovative activities can be adequately addressed with current data sources such as BRDIS and the Community Innovation Survey (CIS), while others may require new or different kinds of data. For example, less is known about how innovation takes place in organizations that are not R&D active, as is often the case in the service sectors.

Questions addressed during the workshop related to this topic included the following:

- What models exist for assessing different kinds of (e.g., non-R&D) innovation?
- What is known about innovators acquiring inventions from external sources?
- How would better information about the division of innovative labor affect understanding of the propensity to innovate?
- How might surveys and alternative data sources be combined to improve the accuracy and interpretability of measures of different kinds and levels of innovation?

EXPANDING SURVEYS TO REVEAL ADDITIONAL INFORMATION ABOUT INNOVATION PROCESSES

Wesley Cohen (Duke University) discussed two prominent survey-based sources of innovation data—the CIS and the Division of Innovative Labor (DoIL) survey, the latter of which he cocreated with Ashish Arora and John Walsh (Arora et al., 2016a). The DoIL survey was designed to shed new light on the key sources of innovation and competitive advantage of American companies.[1] In his presentation, Cohen addressed the question, discussed earlier by Gault and others, about what is meant by innovation and how surveys should be structured to measure it. He also reviewed substantive findings from the DoIL survey.

Cohen reported the top-level questions from the most recent CIS were (1) "Did your enterprise introduce a product innovation?"—meaning one or more new or significantly improved goods—and (2) "Were any of your product innovations new to your market or only to your enterprise?" or a "first in your country, Europe, or the world?" The CIS frames questions about innovation at the firm level. For example, regarding revenues and innovation, it asks what percentage of a firm's total turnover in a given year was from new-to-market (NTM) innovations, new-to-the-firm innovations, and significantly improved products during the 3-year reference period.

The CIS and DoIL surveys generated the following estimates of the share of manufacturing firms introducing innovations for the 2007-2009 period in selected European countries and the United States:

New-to-Firm Innovation
- Germany: 49%

[1]While Arora et al. (2016a) focus on manufacturing, the project collected data from a number of software firms in other service-sector industries for a total of about 5,200 respondents. The response rate was about 30 percent, but the authors collected information on nonrespondents and were able to evaluate response bias in a number of ways.

- UK: 34%
- France: 28%
- U.S.: 42%

New-to-Market Innovation
- Germany: 23%
- UK: 17%
- France: 19%
- U.S.: 16%

Cohen said a key concern with the above statistics is how to interpret what respondents mean by new or significantly improved goods. A "yes" answer could indicate something comparatively trivial such as a new toothpaste formula, or something major, such as the first 3D printer. What respondents have in mind when answering these questions needs to be well understood in order to attach meaning to this kind of information.

Compared with the CIS, the DoIL is less broad in spirit. Its objective was to characterize the contours of what its developers call innovative labor and to emphasize the distinction between invention and innovation. The questions at the core of the survey were intended to determine to what extent innovators—companies introducing something new to the market, or commercializing new products or significantly improved products—acquire inventions from external sources and channels. The framework allows researchers to estimate and compare the value of externally acquired inventions by source—buyers, suppliers, another firm in the industry, universities, and to identify which channels—licensing, contracts, cooperative efforts—and the like—were exploited. Cohen reviewed some of the questions asked in the survey and the reasoning behind them.

In order to identify innovating firms, the DoIL survey asked: "In 2009, have you earned revenue from any new or significantly improved goods or services in [INDUSTRY] introduced since 2007, where 'new' means new to your firm?"

In order to focus on respondents' most important innovations, the line of questioning continues with: "Of all the new or significantly improved products or services you brought to market in [RESPONDENT INDUSTRY] during the 3 years, 2007-2009, think of the one that accounts for the most revenue."

Furthermore, to identify respondents as NTM innovators, the survey asks: "Did you introduce this innovation in your industry before any other company?"

This question structure builds directly on the definition of innovation from the *Oslo Manual* and from the CIS, Cohen pointed out. The major advance created by DoIL is that it identifies a specific line of business for a given respondent. The survey is not for the firm as a whole, but is conducted at the business-unit level.

In addition to NTM innovators, DoIL enables identification of imitators—cases in which something new was introduced to the firm but which was not new to the industry. The question framing allows for calibration regarding what meets the threshold to classify as an innovation for a given purpose. Analysts can set the criteria—for example, the percentage of sales—for what is considered important.

Summary statistics reveal an NTM innovation rate in the United States of about 16 percent for the 2007-2009 period across all manufacturing industries, with innovation occurring disproportionately among large firms (38% versus 23% and 13%, respectively, for medium and small firms). There was also a wide range of innovation rates by industry: At the top end were instruments and electronics, at 37 percent and 33 percent, respectively; at the low end (8-9%) were wood, minerals, and metals. About 27 percent of respondents reported imitating, and these rates were more stable across industries and firm size relative to the innovation rate. The percentage of sales from NTM innovations was also highly skewed by industry (e.g., 31% for medical equipment and 5% for metals). For NTM innovators, the most important new-to-market innovation accounts for the bulk of all new-to-firm sales (about 70%).

Cohen said the survey asks respondents about the sources and channels from which the most important innovations (contributing to the overall design or the development of prototypes or conceptualizations of technology) originated. About 49 percent of respondents reported externally sourced inventions. The most pervasive source of innovations was customers; the most valuable ones originated from technology specialists, which include contractors, universities, and independent inventors. Innovations were acquired through market channels (e.g., licensing, contract, and equity acquisition) and nonmarket channels. The latter accounted for almost two-thirds of the cases, with cooperative efforts at 61 percent.

Returning to the question of where in the spectrum (from trivial to very important) a product innovation falls, Cohen suggested that surveys could do more to capture the relevant dimensions in measurable ways. Supplementary indicators of economic value and technical significance could be developed that permit assessment of the significance of innovations reported by respondents. DoIL indicators along these lines were constructed based on questions asking:

- the percentage of business unit sales due to the focal innovation;
- whether, in order to commercialize a focal innovation, the innovator: (1) developed new sales and distribution channels or (2)

invested in new types of equipment or hired employees with skills different from existing employees; and
- whether the focal innovation is patented by (1) the innovator or (2) an external source.

The last question is less a measure of economic value, since most patented inventions are never commercialized, and more a measure of technical novelty.

Cohen used Figure 3-1 to present findings about the percentage of business unit sales from focal innovation. The vertical axis is the percentage of innovating firms, and the horizontal axis indicates the percentage of sales accounted for by that group of innovating firms. For example, looking at the second-to-last bar on the right, about 8 percent of innovating firms report their most important new or significantly improved product accounted for more than 50 percent of sales.

Findings from the DoIL also indicate levels of investment in commercializing focal innovations (e.g., about 25% of respondents reported buying new equipment, hiring new personnel, and developing new distribution channels), as well as patent rates among NTM innovators (e.g., the rate is about 42% in manufacturing).

Cohen discussed findings about the correspondence between the percentage of sales due to focal innovation and other indicators. As shown in Figure 3-2, the degree to which focal innovations accounted for sales

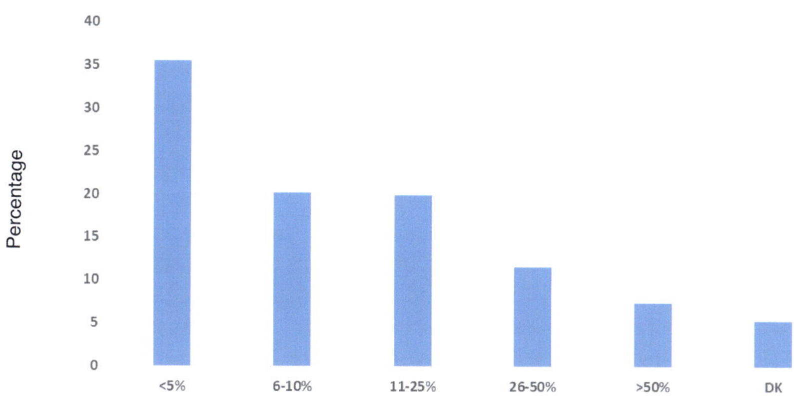

FIGURE 3-1 Percentage of innovating firms by percentage of sales from focal innovation, 2009.
NOTE: N = 1,062 new-to-market innovators; DK = don't know.
SOURCE: Workshop presentation by Wesley Cohen, May 19, 2016; data from Division of Innovative Labor Survey (see Arora et al., 2016a).

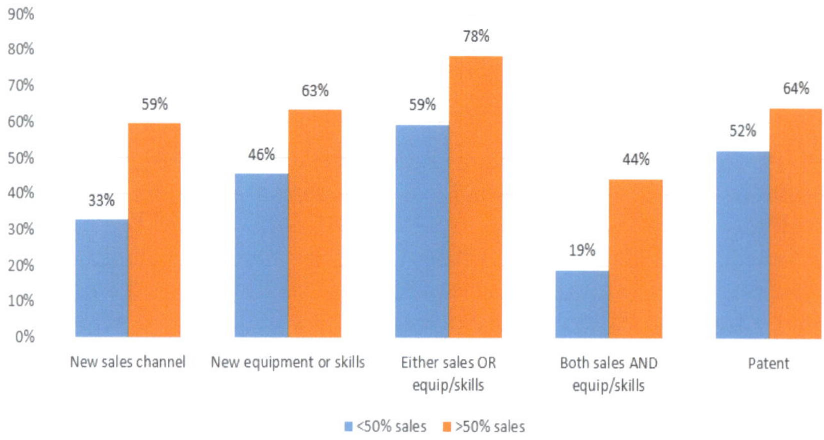

FIGURE 3-2 Correspondence between percentage of sales due to focal innovation and other indicators, 2009.
NOTES: N = 1,062 innovating firms. Y axis is the percentage of innovating firms in each category of the x axis. X axis categorizes other kinds of innovations, further classified by percentage of sales accounted for by the focal innovation (< 50% sales, > 50% sales).
SOURCE: Workshop presentation by Wesley Cohen, May 19, 2016; data from Division of Innovative Labor Survey (Arora et al., 2016a).

among innovating firms (less than or more than 50%) tended to correlate with other metrics. For example, looking at the second-to-last set of bars from the right, among firms whose innovation accounted for more than 50 percent of sales, 44 percent also indicated that they developed new distribution channels and invested in new sales and equipment. These different kinds of innovation are interrelated.

Cohen concluded that the observed high reliance on external sources for invention suggests that to understand drivers of innovation, researchers must consider the extent and implications of the "division of innovative labor." He stated, "It's not just large firms, it's not just startups, it is not just universities." Rather, measurement of the performance of the system as a whole must be focused on the relationships across them. He said his methodological conclusions were that (1) innovation measures focusing on a specific innovation offer accuracy and interpretability, and provide latitude and discretion to an analyst trying to understand various phenomena of interest; and (2) multiple measures tied to a specific innovation can reflect dimensions of economic and technical importance, mitigating ambiguity surrounding such terms as "innovation" and "new or significantly improved."

CHANGING DIVISIONS OF LABOR IN INNOVATION

Ashish Arora (Duke University) reinforced many of Cohen's points, such as that thinking about innovation purely from the point of view of the innovating firm is misleading. Arora said he based his presentation on Simon Kuznets' premise that the paramount feature distinguishing modern economic growth is the systematic application of scientific knowledge to the problem of human production.[2]

Arora made distinctions among invention, innovation, and knowledge. Early in the country's history, for example, innovators in the railroad industry, the technically most sophisticated enterprise of its time, did very little inventing on their own. For the most part, they relied on outside sources and independent invention by employees or contractors. Exemplifying the times, describing the "duties of the patent department" at AT&T, T.D. Lockwood wrote in a 1995 letter:[3]

> I am fully convinced that it has never, is not now, and never will pay commercially, to keep an establishment of professional inventors, or of men whose chief business it is to invent. . . . [T]he duties of the patent department. . . [should be]. . . first and foremost on examining patents or inventions submitted by the public for consideration and second on examining descriptions of inventions forwarded by the company's employees.

Lockwood's inclinations notwithstanding, "industrial research" did emerge, initially to deal with the problem of external invention. The country entered the golden age of American R&D during the 1950s through the 1970s; companies such as AT&T, Dow, DuPont, and IBM worked on a path of innovation-led growth relying largely on internal invention. At about this same time, Arora noted, many statistical agencies began developing their models of innovation measurement, which influenced methodological decisions.

Arora noted that fast-forwarding to more recent years, data from the National Science Foundation (see Figure 3-3) show a return to the pre golden age trend, as the share of privately funded R&D taking place in the economy by large firms has declined.[4]

---

[2]Nobel Prize Lecture, December 11, 1971. Available: http://www.nobelprize.org/nobel_prizes/economic-sciences/laureates/1971/kuznets-lecture.html [August 2016].

[3]Cited by Lamoreaux and Sokoloff (1999, p. 42).

[4]In absolute dollar terms, adjusted using the gross national product (GDP) implicit price deflator, R&D spending by large firms has increased over the period shown in Figure 3-3.

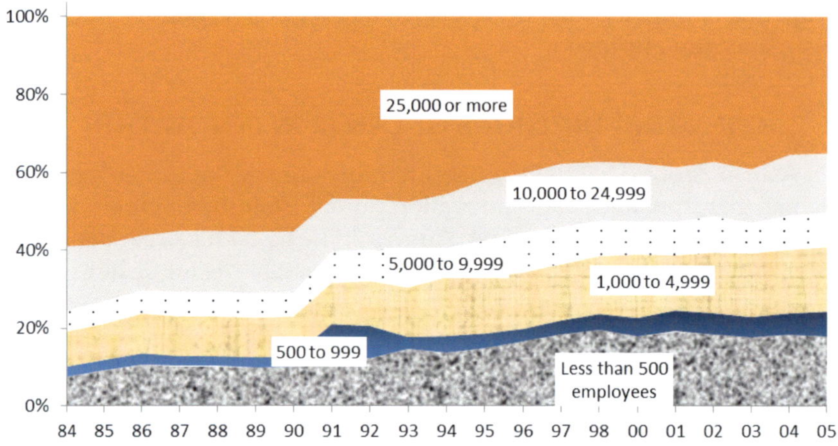

FIGURE 3-3 Share of total privately funded research and development in the United States by firm size, 1984-2005.
SOURCE: Workshop presentation by Ashish Arora, May 19, 2016; data from NCSES, Industrial Research and Development Information System, Historical Data 1953-2007, various tables. Available: http://www.nsf.gov/statistics/iris/search_hist.cfm?indx=11 [August 2016].

Other indicators, such as R&D award winners among Fortune 500 companies, paint a similar picture: large firms—the innovators getting products into the market—are not inventing the inputs themselves at the same rates they were 30-50 years ago.

Arora pointed to an interrelated trend, shown in Figure 3-4, in which the share of basic and applied research as a percentage of total R&D has been in steady decline since the mid-1980s as companies move away from the generation of new knowledge. They are still in the business of using inventions to produce innovations, but, Arora argued, these data indicate a growing separation between knowledge production and the use of that knowledge in invention and then innovation. Given the diminishing role of internal research in large private-sector companies, understanding the mechanisms of external knowledge production and spillovers from universities, startups, and government sources, as well as non-U.S. sources of invention and knowledge, is becoming increasingly important.[5]

Regarding the organizations doing the inventing, in the United States,

[5]Additionally, some companies are moving away from introducing new products and into financial engineering, globalization, and buying businesses (where innovations may be harder to detect and measure).

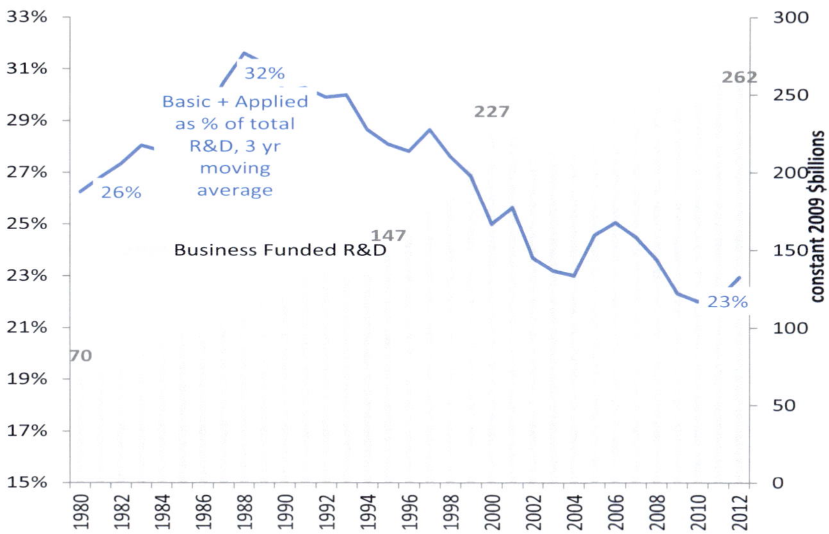

FIGURE 3-4 U.S. business expenditure on basic and applied research, as share of total business funded R&D, 1980-2012.
SOURCE: Workshop presentation by Ashish Aurora, May 19, 2016; data from the National Science Board, *Science and Engineering Indicators 2016*, Appendix Tables 4.6-4.8. Available: https://www.nsf.gov/statistics/2016/nsb20161/#/data [August 2016].

roughly one-half come from sources—suppliers, customers, other firms, consultants, independent inventors, and universities—that are external to the manufacturers who put them to use (Arora et al., 2016b). The channels through which inventions flow to innovators are market-based about one-third of the time. The other two-thirds flow through other informal means, which Arora speculated include public domain, stealing, or some combination. Some are collaborative, cooperative joint ventures. The National Science Board *Science and Engineering Indicators* publications provide information about the extent to which invention originates from science by showing the percentage of patents citing science and engineering literature.[6]

In open discussion, commenting on external sources of inventions (and inventive labor), Alfonso Gambardella (Bocconi University) noted an important distinction between market and nonmarket mechanisms.

---

[6]See, for example, the National Science Board, *Science and Engineering Indicators 2016*, Table 5-30. Available: https://www.nsf.gov/statistics/2016/nsb20161/#/report/chapter-5/outputs-of-s-e-research-publications-and-patents [August 2016].

Current data are geared to capturing the former (e.g., licensing of patents), while researchers have to be more creative with the data to glean information on the latter (e.g., taking advantage of relevant inputs that are in the public domain). The nonmarket-mediated mechanisms are still largely unknown in terms of understanding their role in economies. Cohen agreed that nonmarket mechanisms are important, noting that, for example, cooperative efforts in these sorts of relationships were shown to be extraordinarily high in the Cohen/Arora surveys. Co-invention, he said, probably accounts for the majority of invention, but little systematic data exist on these arrangements at an economy or sector-wide level.

Citations by U.S. patents to science and engineering articles have been trending up from about 16 percent in the early 2000s to well over 20 percent in more recent years. Arora said this provides some indication that innovative activity is not getting separated from the scientific or technical base, as some have argued.

Arora concluded that currently collected data tell a lot about the division of labor in invention, adoption, and innovation. The data currently can show how innovation systems are evolving in the U.S. economy. He suggested more could be done, especially through data linkages used to measure idea flows. He said this requires improved data on the sources of invention, sources of knowledge, and flows of external knowledge.

RESEARCH USING LONGITUDINAL BUSINESS DATA AND OTHER STATISTICAL AGENCY DATA THAT HELP MEASURE INNOVATION

Javier Miranda (U.S. Census Bureau) presented work measuring the high tech sector using administrative records at the Census Bureau. Picking up the discussion thread from Sichel's presentation, Miranda argued that regardless of the productivity measurement issues, other outcomes show that different trends emerged for the high tech sector after 2000. His research (with coauthors John Haltiwanger, Ron Jarmin, and Ryan Decker) indicates a decline in the number of startups in the high tech sector and in the number of jobs generated by these startups (see Figure 3-5). Specifically, the number of startups that become superstar firms that generate a lot of jobs in the economy has declined over the period.

The secular decline in business startup rates over the 1981-2013 period—a 12 percent decline peak to peak between the recessions in the 1980s and the Great Recession—is a concern in that young firms disproportionately contribute to employment and productivity growth. But the answer to whether the economy or a given sector has become more or less innovative depends on what types of startups are in decline. Miranda argued that if it had been "mom and pop" businesses or mainstream

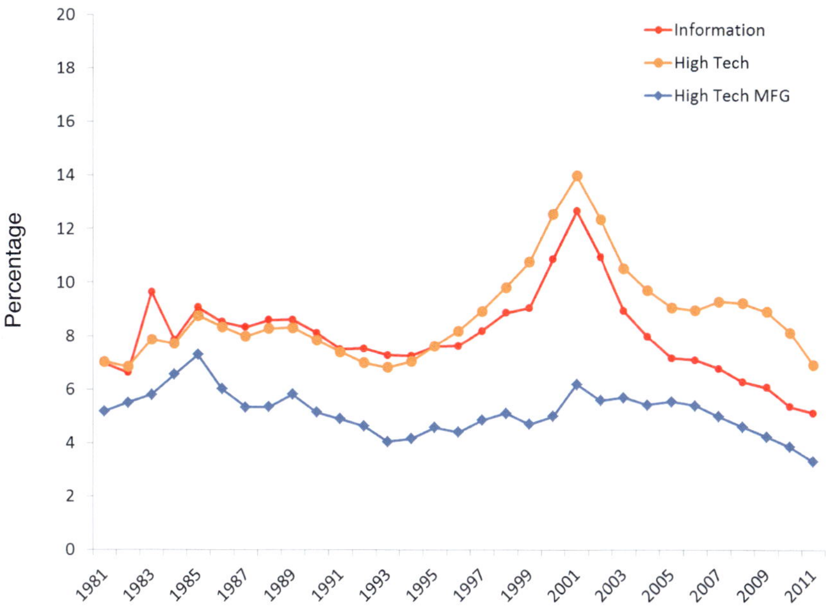

FIGURE 3-5 Share of employment in young firms in high tech sectors.
SOURCE: Workshop presentation by Javier Miranda, May 19, 2016; data from Decker et al. (2015).

businesses that are not very productive, then the decline might be less worrisome. However, if the decline has been in transformational entrepreneurs creating innovative businesses, then it would be a greater concern. Research by Miranda and his colleagues (e.g., Decker et al., 2015) found the decline in startup rates during the 1990s was driven mostly by fewer "mom and pop" businesses starting in the retail sector. After 2000, the trend appears to have been largely driven by a decline in transformational entrepreneurs, the type of businesses that historically have generated a lot of jobs.

Given that researchers do not have flags in the data distinguishing between startups by transformational entrepreneurs and mainstream or "lifestyle" entrepreneurs, they have examined measures of skew in the firm growth distribution. Miranda explained that the idea takes advantage of the fact that relatively few firms become superstars that generate a lot of jobs. In other words, a few innovative firms generate a disproportionate amount of new jobs. For the high tech sector—a composite of industries (e.g., information services, computer and peripheral equipment, software publishers, computer system design, and related services) with more than 25 percent of its workers in science, technology, engineering, and mathematics (STEM)—Haltiwanger et al. (2014) found a high level of dynamism during the 1990s: entry of new firms coinciding with U.S. productivity growth. Around the time of the "dotcom bust" and the accession of China to the World Trade Organization, a sharp reversal and decline in high tech startup activity took place. Likewise, as shown in Figure 3-5, the decline in the share of employment in high tech jobs accounted for by young firms began in 2000 and, by 2011, it was about half of what it had been a decade earlier. The decline coincides with a slowdown in productivity growth as measured using official statistics (or, Miranda pointed out, using Fernald's numbers cited by Sichel above).

Miranda focused on the firms that generated a lot of jobs, which he said requires looking beyond R&D data. He said measures of trends in STEM worker employment are used to help complete the picture. The research shows the importance of looking at measures of dispersion and skewness to understand what is happening in this sector. Statistical agencies are looking into different data products to get at this type of heterogeneity.

Kristin McCue (U.S. Census Bureau) reported on two emerging developments at the Census Bureau relevant to the study of innovation. The first involves enhancements to the Business Dynamics Statistics (BDS) Program that will provide useful information on innovative businesses; the second involves two new surveys of small, young firms designed to collect information about innovation that is not available from administrative records.

The extensions envisioned for the BDS Program—which produces data on job creation and destruction, establishment openings and closings, and numbers of startups and shutdowns, as well as the age and size of firms—is to produce a high tech industry measure based on the work of Hecker (2005). In the near term, statistics will be produced for a grouping of a set of industries with disproportionate STEM employment using the definition identified above by Miranda. Longer term, statistics may be produced by 4-digit North American Industrial Classification System (NAICS) codes so users can create their own groupings of high tech industries, she said.

The BDS Program is also looking to produce statistics on high-growth employers. The idea is to generate data on percentiles of employment-weighted growth distribution in order to be able to track the skewedness of the distribution over time and across sectors. There have also been efforts to create statistics on whether businesses and employees have been involved in patenting activity. This involves using the linked employer-employee data to link inventors to their data in the census. For all of these products, the underlying linked microdata will be potentially accessible to researchers in the Census Bureau's Federal Statistical Research Data Centers, she said.

Two new surveys will derive content from the NCSES Microbusiness Innovation Science and Technology Survey (MIST). One, the Micro BRDIS, is being developed by NCSES. The other, the Annual Survey of Entrepreneurs, includes modules of eight questions from the MIST Survey covering process and product innovation and R&D costs, funding, purchases, and employees. These are both surveys of small and young firms. As in past collaborations between agencies, such as between NCSES and Bureau of Economic Analysis in developing new data for adding R&D and intangibles into the National Income and Product Accounts, questions of data confidentiality are regularly dealt with.

DESIGN INNOVATION AS AN ALTERNATIVE OR COMPLEMENT TO R&D

Bruce Tether (University of Manchester) focused on design in his presentation, primarily "industrial design" as opposed to engineering design. He observed the latter is often thought of as an activity taking place within R&D, while the former typically is not. Tether cited a number of prominent examples of new product introductions driven by design: Ford's Model T was replaced by the Model A in large part because people wanted a new, more varied product. There was nothing particularly innovative about it from a technological perspective. Similarly, Apple introduced the iMac computer with the intention of making computing fun and accessible: the innovation was in the presentation and user-interface, rather than in the underlying technology. In both cases, Tether said, understanding and responding to user needs and desires was central. Design also applies to services. For example, one way that Virgin Atlantic competes with larger airlines is by investing heavily in design to create a different customer experience.

So what, Tether asked, is design? He noted the term is as malleable as "innovation." The traditional view is that design is something applied after difficult complex engineering decisions have been made, when the goal is to make a product look attractive: in this view, it is about outward appearance. Even if this were the extent of it, Tether argued that design matters. Its growing importance is indicated by a steady rise in the number of design patents issued since the mid-1980s, as shown in Figure 3-6: more than 20,000 design patents annually since 2006. The total number of patents has also risen substantially, so, as a share of patents, the share of

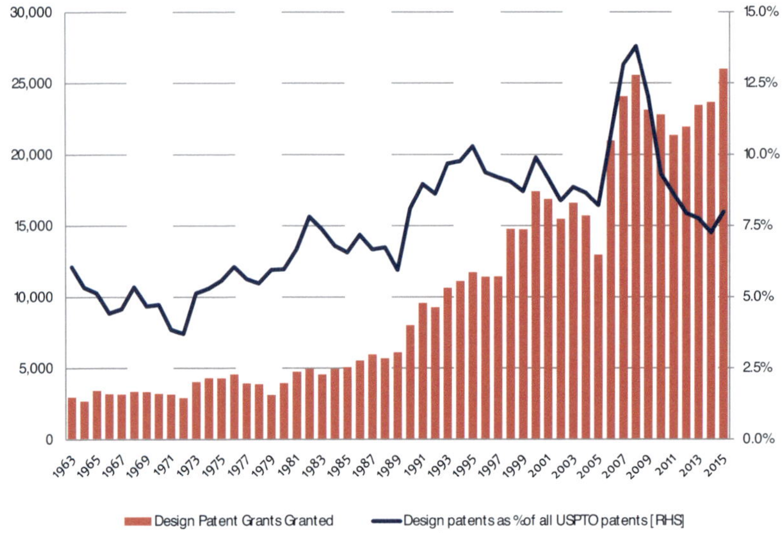

FIGURE 3-6 U.S. Patent and Trademark Office design patents granted, 1963-2015. SOURCE: Workshop presentation by Bruce Tether, May 19, 2016; data from U.S. Patent and Trademark Office, Patent Technology Monitoring Team. Available: http://www.uspto.gov/web/offices/ac/ido/oeip/taf/us_stat.htm [August 2016].

design patents has only gone up marginally. Nonetheless, these data give some indication of the significance of appearance.

The nature of the big design patent-owning firms (e.g., Apple, Honda, LG, Microsoft, Nike, Procter & Gamble, and Samsung) indicates that there is a relationship with technologies, such that design goes beyond aesthetics. These companies have both utility and design patents. The modern view of design is that it brings together creativity through insights and integration of ideas. Tether suggested a good example is IDEO, a global design firm, which created a program for Bank of America called "Keep the Change" to entice people to save more (every time customers buy something with their debit card, the bank rounds up the purchase to the nearest dollar and transfers the difference from their checking into a savings account). This kind of innovation required a deep understanding of people's needs and wants and how to realize them.

Tether provided statistics based on an Innobarometer survey, conducted in 2015, about the extent to which firms in various countries are

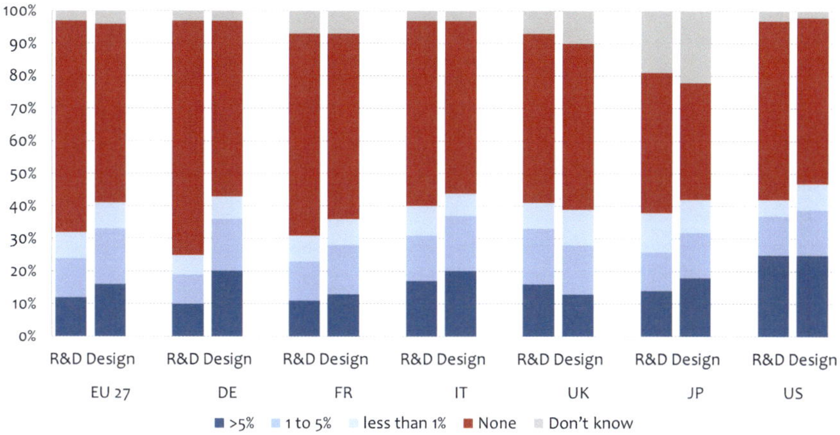

FIGURE 3-7 Investments in research and development (R&D) and design (outside of R&D), by country and intensity.
NOTES: "Intensity" refers to the share of turnover (or revenues) that is spent on R&D and design. The Y axis is the percentage of companies investing in R&D or design in 27 European Union countries (EU 27), Germany (DE), France (FR), Italy (IT), United Kingdom (UK), Japan (JP), and the United States (US).
SOURCE: Workshop presentation by Bruce Tether, May 19, 2016; data from Innobarometer 2015 (Montresor and Vezzani, 2015).

engaged in design and R&D.[7] As shown in Figure 3-7, more than 40 percent of U.S. companies invested in design outside of R&D, a slightly higher share than the share of firms that invest in R&D. Similar findings were found for the EU-27 and for Japan.

Tether described how it is not only how much firms spend on design that matters but also the role of design in the firm. The Danish Design Centre developed a design ladder to indicate this. Firms can use design as a strategy, where it is a central feature in the business concept; as a process, where it is integrated early in product development; or as a last finish, to give the product form. Finally, he acknowledged, some firms do not use design at all.

Controlling for various factors—such as investments in R&D, investments in software, branding, the age and size of companies, sector, and country—research has found that firms that spend on design are

[7]Innobarometer is an annual survey sponsored by the European Commission on activities and attitudes related to innovation. Available: http://ec.europa.eu/growth/industry/innovation/facts-figures/innobarometer/ [August 2016].

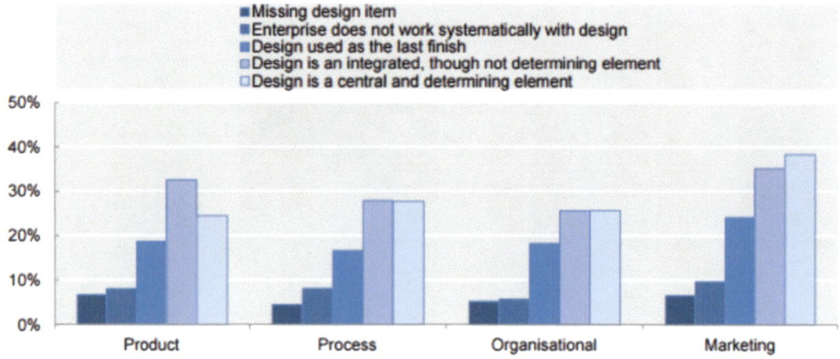

FIGURE 3-8 Marginal effects of use of design on the probability of introducing an innovation, 2010.
NOTE: Marginal effects obtained from probit model estimations, controlling for size and sector. Baseline = "don't know/not relevant." All coefficients are significant at the 5% level.
SOURCE: Workshop presentation by Bruce Tether, May 19, 2016; estimates from Galindo-Rueda and Millot (2015), using data from Statistics Denmark, based on Community Innovation Survey 2010 results for Denmark.

more likely to innovate and have higher sales from innovation. Work by Galindo-Rueda and Millot (2015) using the Danish CIS data reveal an increasing impact from using design more strategically. As shown in Figure 3-8, firms that use design in a centralized way are more likely to introduce innovations than those that do not use it or use it as add-on styling.

Another recent study using the Innobarometer survey (Montresor and Vezzani, 2015) found similar marginal effects on whether or not firms introduce product or service innovations in terms of the centrality of design. As shown in Figure 3-9, when design becomes more central, the probability of introducing innovations increases. There is also a marginal association with sales, although not necessarily a causal one. Again, where design is a central element, the share of sales linked to product/service innovations is higher.

Tether summarized his presentation with the following points:

- Design has mattered as a competitive weapon for a long time, and yet it has been on the fringes of innovation measurement and, as a result, is not well understood.
- Recent studies show that spending on design, and its role in the firm, affects the introduction of innovations. It is not adequate to measure just the spending on this input, it is also important to know where design is positioned in the firm.

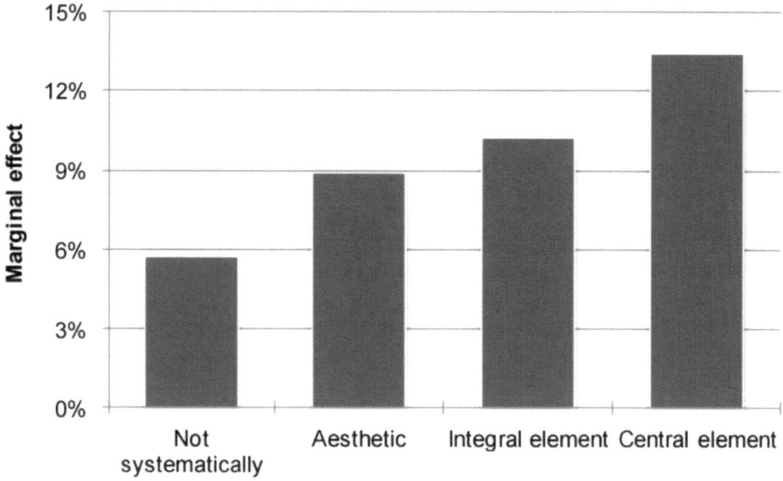

FIGURE 3-9 Marginal effects of use of design on the probability of introducing product/service innovations.
SOURCE: Workshop presentation by Bruce Tether, May 19, 2016; estimates from Montresor and Vezzani (2015), using data from Innobarometer 2015.

- Design is increasingly seen as part of a process that integrates an understanding of users' needs, which are not always explicit. Design relates to matters of appearance, and appearance matters (people make selection choices on the basis of how things look), but design now goes well beyond that.
- Design varies across different sectors. A lot of attention is focused on the high tech sectors, but that is only a small part of the economy. Design is important in low tech sectors and, increasingly, in services.
- Among the challenges in measurement are (1) there is no agreed-upon standard for defining design;[8] (2) design is often hidden: two companies may invest the same amount, but this may not appear to be the case if one has design activity subsumed into R&D and marketing; (3) beyond the amount spent, the arrangement of the spending is also important—if spending is fragmented, it seems to have less impact than if it is concentrated;

[8]*Guidelines for Collecting and Interpreting Design Data—A Proposal for a Future Barcelona Manual on Design* prepared for the European Commission, June 2014, made some proposals for a standard. Available: http://www.bcd.es/site/unitFiles/4242/measuringdesignvalue.pdf [August 2016].

(4) the quality of design inputs (including key design staff) may matter more than the extent of investment; and (5) contingencies and complementarities are involved, as is usual in innovation studies, that determine when design is likely to be most effective.

AUTOMATION INNOVATIONS

Rob Seamans (President's Council of Economic Advisors [CEA], on loan from New York University) presented on the need to track automation and the use of robotics across sectors of the economy in a way that informs employment and other policies. This dimension is crucial given the relationship between automation in manufacturing and services and employment and wages.

Seamans touched on a number of topics from the *2016 Economic Report of the President*, which included a section on robotics and automation.[9] Extensive evidence exists that automation leads to productivity growth (see, e.g., Bloom et al., 2012), but there has been little research on the robotics component of automation. Seamans cited a paper by Graetz and Michaels (2015) quantifying the shipment of robotics into different sectors of the economy in different countries from 1993 to 2007. This research finds that use of robotics led to, on average, a 0.36 percentage point increase in a country's annual labor productivity growth, accounting for slightly more than one-tenth of overall GDP growth during that time. Although small, Seamans pointed out that its impact is similar in magnitude to that resulting from steam engines in the United Kingdom in the late 19th century—and about one-tenth of the productivity growth that occurred during this time period.

Robotics can either complement or substitute for labor. While macroeconomists tend to focus on top-line productivity growth, according to Seamans it is also important to know whether the use of robotics comes at the expense of labor, and, if so, whether policies exist to mitigate this potential downside. Seamans highlighted a few examples, including China's Great Wall Motors Assembly Line, where robots have replaced workers who previously riveted parts onto cars. More generally, a large share of robotic shipments go to the automotive industry in such countries as China, Germany, and the United States, and most notably Japan, where there are high ratios of robots to workers. In this case, there might be more substitution than complementing, although, even then, the car robots are likely increasing the productivity of workers at higher plant managing levels. Another example that Seamans pointed to is Amazon's investment

---

[9]Available: https://www.whitehouse.gov/administration/eop/cea/economic-report-of-the-President 2016 [August 2016].

in a company called Kiva Systems, which makes robots that take items to pickers located in a central location in the warehouse. These machines traverse aisles, locate items, and bring them to workers who then perform the higher-level task of figuring out what they need to pick in order to fill the order. These robots are performing an essentially complementary role that increases the productivity of the human pickers and order processors.

Data from the International Federation for Robotics (IFR) indicate that total annual shipments of industrial robotics began to increase rapidly around 2010, and they have nearly doubled since then.[10] Similarly, beginning around 2011-2012, there was a large uptick in the number and rate of patents that cite the robot class. These trends suggest, according to Seamans, that the 0.36 percentage point contribution to productivity growth estimated by Graetz and Michaels (2015), which only included data up to 2007, may significantly understate current contributions.

The effects of automation on labor are likely to be geographically heterogeneous, given that automation affects industries differently and that some industries are more geographically concentrated than others. To illustrate this point, Seamans presented the number and share of workers in each county who work in industries receiving the most robotics shipments. Not surprisingly, large counties such as Los Angeles and Santa Clara have high numbers of potentially affected workers. In terms of the *share* of workers, counties in which the dominant employer is concentrated in a highly automated industry are at the top of the list. In Calhoun County, Arkansas (the home of three large aerospace and defense firms—Aerojet, General Dynamics, and Lockheed Martin), close to 100 percent of the workforce is potentially affected. Rates almost this high are also obtained in Bristol Bay, Alaska, which is dominated by canning and seafood processing, and Saluda County, South Carolina, home to large poultry and meat processing plants. From a policy maker's perspective, Seamans said, these locations are interesting because robotics and automation are having large impacts (positive for some workers, negative for others) in the local economy.

Researchers have also examined the link between the occupational "threat of automation" and wages across industries. Data from Frey and Osborne (2013) show the probability of automation for each of about 700 types of occupations. BLS data on wages can be matched for each of those occupations, as shown in Figure 3-10. In the figure, each dot represents an occupation. Toward the left-hand side of the graph are occupations such as podiatry, civil engineering, the clergy, interior design, and human resources management, where jobs require skills such as manual dexterity

[10]Sander and Wolfgang (2014) have estimated that worldwide spending on robotics—$26.9 billion in 2015—will rise to $66.9 billion by 2025.

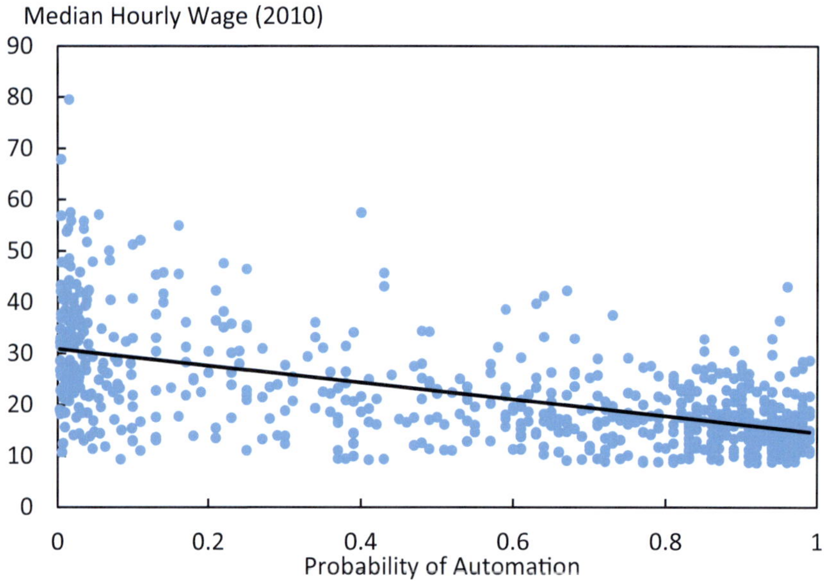

FIGURE 3-10 Median hourly wage (2010) and probability of automation by occupation.
SOURCE: Workshop presentation by Rob Seamans, May 19, 2016; wage data from Bureau of Labor Statistics; probability of automation estimates from Frey and Osborne (2013); and President's Council of Economic Advisors' calculations.

or complex thinking that do not lend themselves easily to automation. On the right-hand side are occupations such as tax preparation, telemarketing, and cashier and office work where less manual dexterity is required or where there is more automatable thinking. Occupations that are easier to automate than others appear to have lower wages associated with them.

Seamans concluded that the topic warrants more research attention, given the recent uptick in robotics (whether measured in units, revenues, or patents), the potential impact on productivity growth and labor displacement, and the geographic diversity of these trends. Data on robotics shipments are available from the IFR at the country-industry-year level. There are also data on the probability of automation, produced by Frey and Osborne, McKinsey Global Institute, and others, as well as patent application data, particularly patents in the "robot" class. Seamans argued that none of these measures is at a granular enough level to provide information to help policy makers track how automation is affecting industries and economies at a local level. Thus, in order to facilitate research on some of these important relationships, he suggested that NCSES could prioritize developing more detailed data indicating which industries are using robots, how much they are spending, and the extent to which robots are substituting or complementing workers. One possibility would be to systematically add questions to a U.S. survey along the lines of "Last year, how much money did your establishment spend on robotics?" and "Has your establishment considered using robotics instead of human labor?"

In discussion, Sheryl Winston Smith (Temple University) commented that robotics is an important input into machine learning; it is essentially a complement in production, but one that is not going to show up in robotics patenting data sources that would attribute the contribution to other categories. Seamans added it is the responsibility of policy makers to track affected industries and geographies, to anticipate the downstream effects of these innovations (e.g., they may boost productivity, but with potentially adverse effects on labor), and to consider appropriate policy responses.

4

The Role of Individuals (and Networks of Individuals) in Innovation

The workshop devoted a session on the role of individuals and teams in the innovative process, which has been less studied than the role of institutions and organizations. How people's educations, entrepreneurial talents, and other human capital characteristics result in innovation are vast and largely unresolved research areas. Linkages among individuals and between institutions are also important to consider.

KNOWLEDGE DIFFUSION, EMPLOYEE MOBILITY, AND ENTREPRENEURSHIP

Rajshree Agarwal (University of Maryland) spoke on improving understanding of how new product innovation interacts with the career profiles of individuals who create them. She commented that entrepreneurship is not a destination, but a step in a longer career lifecycle, and thus the effect of entrepreneurial firm fates on individual career lifecycles needs more attention if it is to be fully understood. In her view, future research would benefit from a focus on human capital *markets,* which requires combining demand and supply factors to examine life-cycle choices and also considering selection effects as they relate to optimal allocation and reallocation of talent. She also suggested the merging of individual-level career and knowledge innovation datasets to identify systematic sources of bias and to answer questions about what kinds of people firms hire, keep, let go, and how these decisions affect career profiles.

Agarwal observed that few studies have investigated how entrepreneurship affects long-term career lifecycles (beyond the new venture directly). There are serial entrepreneurs, individuals who choose to leave and then return to a firm as an employee, and founding teams composed of individuals. One key research question is how the fates of new ventures affect long-term career outcomes of entrepreneurs and founding team members.

Research often focuses on the product market, where primacy is given to demand conditions, and on individual career choices, where primacy is given to preferences and incentives. A singular focus on either one of these ignores the fact that human capital markets need to clear. More work is needed to understand the role of mobility and entrepreneurship in the allocation and reallocation of talent. Agarwal noted that several factors, on both the demand side and the supply side, impact mobility and entrepreneurship. There are protection mechanisms and costs associated with mobility across organizations, even startups. Mobility and entrepreneurship are often impacted in the same direction, but sometimes the effects are divergent. For example, collusion among high tech firms to not poach talent may reduce mobility, but it may enhance entrepreneurship. Little is known about how other factors (listed in Table 4-1) impact mobility and entrepreneurship. For example, she queried, how does family composi-

TABLE 4-1 Factors Affecting Mobility and Entrepreneurship in Human Capital Markets

Demand Side	Supply Side
Protection Mechanisms • Noncompetes • IP protection	Mobility Costs • Family ties • Location preferences • Health care and other benefits
Collusion/Thin Markets Competition versus cooperation in firm interactions	Information Asymmetries Knowledge contexts—entrepreneurship by users, employees, and academies
Firm Specificity/Complementarities • Regional policy impacting knowledge flows	Individual Preferences for Job Attributes • Security versus growth/risk preferences • Social support programs
Social Complexity • Team-embedded knowledge • Technological complexity • Regional clustering of knowledge	Individual Preferences for/against Entrepreneurship • Taste for autonomy, mastery, purpose

SOURCE: Workshop presentation by Rajshree Agarwal, May 19, 2016.

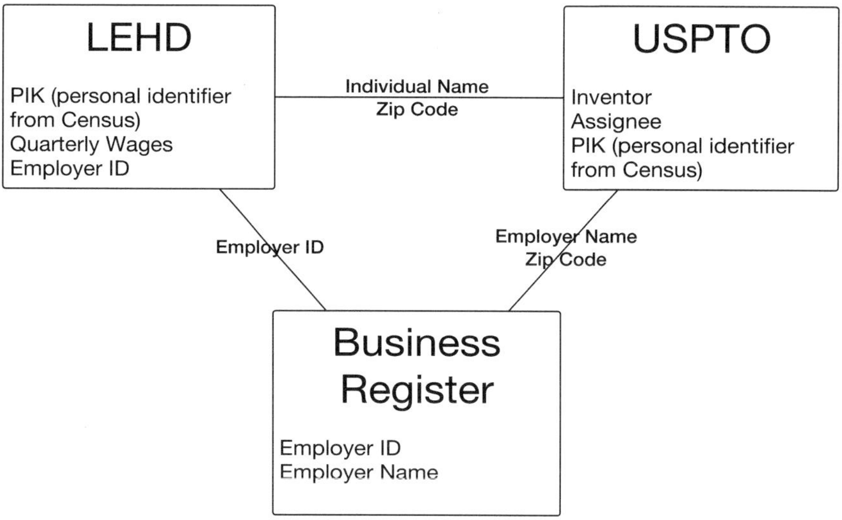

FIGURE 4-1 Linking census and patent data (see Graham et al., 2015).
NOTES: LEHD = the Census Bureau's Longitudinal Employer-Household Dynamics Program (see http://lehd.ces.census.gov/ [August 2016]). USPTO = U.S. Patent and Trademark Office (see http://www.uspto.gov/patent [August 2016]). The Census Bureau's Business Register is described at https://www.census.gov/econ/overview/mu0600.html (August 2016).
SOURCE: Workshop presentation by Rajshree Agarwal, May 19, 2016.

tion affect individuals' mobility and entrepreneurship decisions, and the associated wage outcomes? Where do other social processes factor in?

Agarwal argued that mobility, entrepreneurship, and innovation datasets need to be linked to study how entrepreneurship and innovation relate to career lifecycles. Figure 4-1 provides an example of the kind of linkages she said would be helpful in answering a range of important questions, such as:

- What types of bias impact patent-based measures of mobility?
- Are inventors more productive when they take co-workers with them?
- Is noninventor mobility a source of knowledge diffusion?
- How much are inventors able to appropriate from a patent their employer owns? Do they see long-run wage/career impacts?
- Is there evidence that inventors are "stealing" ideas?
- Are inventors filing patents soon after they leave for a startup?

MOVING BEYOND PATENTS TO EXPLOIT NEW DATA AND METHODS TO STUDY INNOVATION

Lee Fleming (University of California, Berkeley) continued the discussion of patent data, new metrics, and data linkages by asking the question, "How can we be more clever in using our data?" He offered several observations to motivate this question:

- Patent data are overused and abused; they are very easy to observe, but not necessarily the right thing.
- Researchers need to stop relying solely on patent counts and citations to measure innovation—richer measures are not that hard to calculate.
- Advances in machine learning and natural language processing are useful, though they need thoughtful application.
- Newly available data and tools provide opportunity to advance understanding of innovation.

In developing these themes, Fleming posed several research questions, the first of which was what happens when an individual changes fields. One viewpoint, standard among economists, is that a person gives up expertise in his or her old field, which is typically harmful to earnings. A second viewpoint, emphasized by Thomas Kuhn (1962, pp. 89-90), is that "almost always the men who achieve these fundamental inventions of a new paradigm have been either very young or very new to the field whose paradigm they change."

Fleming argued that to answer his question, novelty must be separated from value. Some products may be quite novel yet provide relatively little in terms of market value. He hypothesized that the patents invented by those who change fields are likely on average to be more novel and less valuable. One alternative for testing this hypothesis is to use citations—but they have been used to measure value *and* novelty, so this is unlikely to work; another alternative is to use a machine learning technique that enables hundreds of papers to be traced to inventions and inventors.

Easily accessible data on patent renewals can be used to create measures of value. But, Fleming asked, what about novelty? Looking back through the patent corpus, one can identify where key words first appear and weight them by their future use. This can be done, for example, with terms like "nontransitory," a limitation put in mainly in the software patent field, and other important words, such as "browser and computer executable," "http," "java," and so on. Combining information on new words that show up in patents with person-level data, it is possible to demonstrate that new entrants are more likely to invent patents with new words, and less likely to invent patents that are renewed. This kind of research question would be very difficult to answer using only traditional kinds of data and analytic methods.

Fleming next posed a research question about how governance influences innovation. The decade of the 2000s provided a fertile case study period for considering this question because governance was increased (via legislation such as the Sarbanes-Oxley Act of 2002, for example) that forced firms to adopt more independent oversight. Stronger governance could increase innovation, because it requires greater focus and effort; on the other hand, it could inhibit creativity and risk-taking. In the finance literature, Fleming stated, patents, patent counts, and patent citations have not been utilized very successfully to measure this question.

Research by Balsmeier et al. (forthcoming) using Sarbanes-Oxley as an instrument finds that the signals are clear: Increased governance makes firms more productive by some measures—for example, they get more patents, although these tend to be in areas of technology where they have previously been patenting. But, the authors found, increased governance appears to lead to cutbacks on exploration and willingness to try new things—fewer patents are issued to firms in new technology classes.

Analyses relying purely on counts and citations miss the kind of details described above, Fleming asserted. The advance by him and his collaborators came with the realization that an instrument could be developed using a principal components analysis to break out two components of innovative activity: exploitation and exploration. Exploitation captures portfolio measures of firms patenting in known classes, staying close to their previous technological proximity (and inventors are getting older). Exploration captures firms entering new patent classes (and inventors are getting younger). The measure subsequently produced in this research allows for a much deeper and richer look at firms' patenting portfolios than would be allowed by simply counting patents.

Looking at all of the firms in their study (see Figure 4-2), on average, exploration appears to slow considerably over the firms' lifetimes. They inevitably evolve toward patterns of more exploitation as they age.

A third research question posed by Fleming was whether changes in innovation components in the larger economy and the business cycle can be seen. He noted the literature indicating that during downturns in the business cycle, there should be more innovation. The reasoning is that firms are busy making money during the upturns, which increases the opportunity cost of taking risks and exploration. However, analyzing research and development (R&D) and patenting data, the signals are not clear. In Fleming's view, greater detail is required to discern more definitively that exploration goes up in a downturn and exploitation becomes stronger during an upturn.

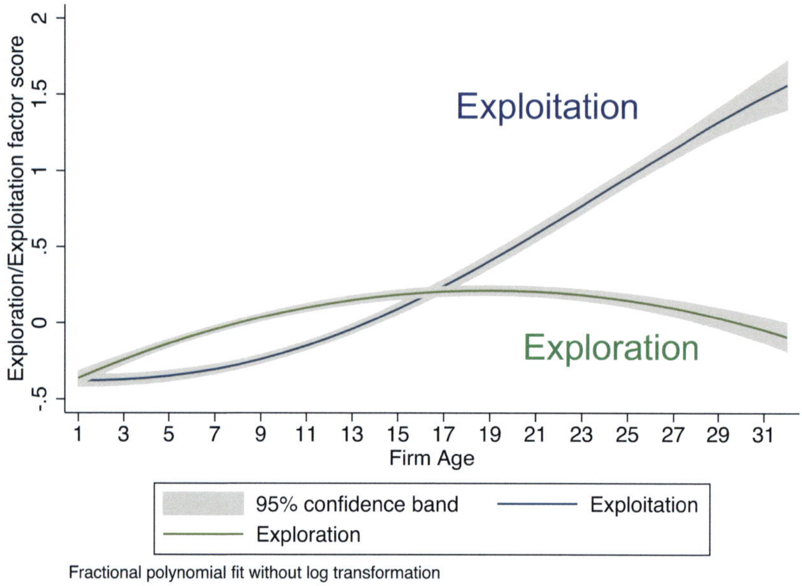

FIGURE 4-2 Exploration/exploitation path dependence.
SOURCE: Workshop presentation by Lee Fleming, May 19, 2016.

Fleming concluded that (1) data on patent counts are (almost) dead as a means to measure innovation; (2) advances in machine learning and natural language processing are valuable for this purpose, although their use requires thoughtful application—crowdfunding is an example of a fertile area; and (3) newly available data and tools provide opportunity—for example, studying real-time inventor networks or exploiting natural language processing to see parts of the innovation economy that are not visible through patenting—particularly when analyzed in collaboration with computer scientists.

Alfonso Gambardella (Bocconi University) discussed the relationship between managerial practices and incentives for research and innovation, and what could be learned from experiments, from large and systematic computerized databases, and from surveys. Researchers have known for a long time that managerial practices are important in terms of performance and productivity (e.g., Bertrand and Schoar, 2003; Bloom and Van Reenen, 2007).[1] Less is known specifically about the impact of manage-

[1]Bertrand and Schoar (2003) showed that managerial fixed effects matter for performance of firms. Bloom and Van Reenen (2007) found that a survey-based measure of management

ment on innovation. Garbardella referred to preliminary work by Manso (2011), Azoulay et al. (2011), and Ederer and Manso (2013), which suggests the relationship is also important. Consistent with Fleming's observation about the business cycle, Gambardella explained that Manso found evidence that short-term rewards and penalties discourage exploration. However, managerial practices have the capacity to offset this by creating an environment that tolerates short-term failures and favors long-term compensation schemes, which suggests that special managerial practices are needed to nurture exploration as opposed to exploitation. Manso and colleagues argue that most pay-for-performance schemes are not ideal for promoting creative work because they penalize short-term failures and favor short-term success. Implementing schemes for a CEO may not be conceptually difficult—compensation can be linked to longer-term performance of the firm. But for midrange employees such as researchers, who actually produce the innovations, it is harder to create verifiable measures of their long-term performance—variables such as stock performance of the firm are too distantly related. He said other instruments and practices are needed.

Gambardella noted that very little is known about what managerial practices are effective for promoting innovation, in part because of some fundamental conceptual hurdles to overcome. It is well established that creative people like independence (Gagné and Deci, 2005; Bartling et al., 2014) and dislike pay for performance (Amabile, 1996). For example, scientists "pay to be scientists" (Stern, 2004), in the sense of accepting lower paying jobs that promise scientific work, although to a varied extent (Sauermann and Roach, 2014). It is also known that motivation matters for innovation (Sauermann and Cohen, 2010). Gambardella said findings from this literature are suggestive that policies about the independence and autonomy of individuals are candidate tools for improving innovation performance at the level of managerial practice. Nicola Lacetera (2009), for example, found that autonomy is a powerful device for increasing scientists' incentives to supply productive effort, especially when their objectives and priorities are not aligned with the top management. Using the National Center for Science and Engineering Statistics (NCSES) Scientists and Engineers Statistical Data System (SESTAT),[2] Sauermann and Stephan (2013) found that 61 percent of respondents in scientific and engineering industries value independence highly versus 81 percent in academia.

Given that inventors value autonomy, Gambardella et al. (2015) argue

practice explains 10 to 23 percent of the interquartile difference in total factor productivity across firms.

[2]See https://www.nsf.gov/statistics/sestat/ [August 2016].

that firms can use autonomy as a tool to motivate their employees, especially when output cannot be used to measure innovation performance. Data from the PatVal European patent inventor survey (Torrisi et al., 2016) provide consistent evidence. In particular, Gambardella et al. (2013) show that firms provide more autonomy to inventors on projects in which they are less motivated.

To better understand the role of management practices in innovation, Gambardella concluded that more data are needed that follow cohorts of workers at science and engineering (S&E) firms so that the role of practices related to tolerance for failure, long-term versus short-term rewards, and autonomy on innovation and performance can be assessed. Scenario-based experiments—for example, asking informed parties how they would respond under different scenarios—and field experiments are also needed.

MEASURING FLOWS OF HUMAN CAPITAL TO FIRMS AND THE ROLE OF UNIVERSITY ADMINISTRATIVE DATA

Paula Stephan (Georgia State University) spoke about measuring the flows of highly skilled individuals to firms. It is common for researchers to track the contribution of universities to innovation in terms of patents, licenses, and startups; it is less common to focus on the placement of highly trained individuals with firms, despite the fact that individuals are a powerful way of transmitting information, especially tacit knowledge. She quoted J. Robert Oppenheimer, who wrote in 1948, "The best way to send information is to wrap it up in a person."

While difficult in the past, it is now possible to identify the placements of highly trained people, such as Ph.D.s, with individual firms by matching university-maintained records and U.S. census data. The approach was demonstrated in a "proof of concept" paper in *Science* (Zolas et al., 2015). Under strict confidentiality protocols, researchers matched data from UMETRICS to census data on employers for a sample of recent Ph.D. graduates from eight universities (Indiana, Iowa, Michigan, Minnesota, Ohio State, Purdue, Penn State, and Wisconsin) who had been supported by grants as students.[3] Linking student grant data with data on dissertations (ProQuest) and census records, the researchers were able to track placements of 1,983 recent Ph.D.s. Additional linkages could, in

[3]The UMETRICS (Universities: Measuring the Impacts of Research on Innovation, Competitiveness, and Science) project collected administrative data on federal research funding and private funding (on some campuses) for 11 universities. In January 2015, the Institute for Research on Innovation and Science (IRIS) was established at the University of Michigan to manage and expand UMETRICS; see http://iris.isr.umich.edu/about/ [August 2016].

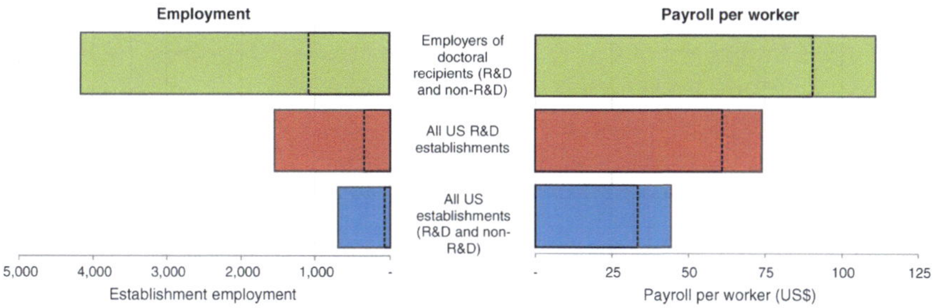

FIGURE 4-3 Doctoral recipients from eight universities (top set of bars) are employed at establishments that are larger and have higher payrolls per worker compared with other establishments.
NOTE: Medians = dashed (inner) lines; means = solid (outer) lines
SOURCE: Workshop presentation by Paula Stephan, May 19, 2016; data from Zolas et al. (2015).

principle, be made. For example, matching the NCSES Survey of Earned Doctorates with census data could allow stay rates of graduates in firms to be estimated.

Results indicate that recently minted Ph.D.s who go into industry tend to be employed at larger, high-wage firms. As shown in Figure 4-3, the distribution of payroll per worker at the establishments where these graduates work is much more skewed than it is for all U.S. establishments or R&D establishments. The placement of doctoral recipients supported by grants also varies by field. As expected, engineers are most likely to go into industry, followed by math and computer scientists; only a very small percentage go to young firms. The data also permit annual earnings distributions to be estimated (from Unemployment Insurance Earnings Records) and compared across government, academic, and industry sectors (industry has the highest earnings), or across disciplines (engineering is highest).

Stephan observed that these kinds of data-linking projects open up a wide range of research opportunities—for example, they allow modeling of how knowledge stocks embedded in human capital contribute to innovation and performance at the establishment level. Research can examine such questions as how different types of support received by Ph.D. students relate to employment outcomes and the role in job placement of social networks between universities and establishments, or across firms and establishments.

Stephan also noted the importance in the global economy of measuring the international mobility of highly trained individuals. Research suggests that internationally mobile scientists and engineers contribute disproportionately to innovation. But, she cautioned, very little is known about how various outcomes are impacted by international mobility. For the United States, the NCSES Survey of Doctorate Recipients (SDR) can be used to follow Ph.D.s who stay in the country by field. A major effort is now under way to follow individuals who receive Ph.D.s in the United States and move, but she said it is much more difficult to follow individuals who arrive for postdoctoral training with a Ph.D. in hand. It is also extremely difficult to compare mobility of scientists and engineers in the United States with patterns outside the United States. Virtually no country has data on emigrant scientists, and there is little empirical evidence to compare the performance of mobile scientists across countries. This, Stephan suggested, is an area where improvement of data should be a priority.

To begin addressing this data gap, Stephan (along with colleagues Chiara Franzoni and Giuseppe Scellato) developed a survey of Italian scientists in four fields who had migrated to 16 countries. Details of the survey, called GlobSci, are summarized in Van Noorden (2012). Stephan said the major advantages of the survey over existing alternatives include the following:

- It is possible to track mobile researchers who returned to their country of origin (if included in the 16 core countries) or who emigrated to a core country.
- Data on "entry point" of foreign born (e.g., Ph.D., postdoc, faculty) are captured.
- Numerous individual-level controls are in place.
- Bibliometric measures of focus articles are generated.

Limitations include a lack of data for China and South Korea and that the questionnaire covers only four fields and provides a snapshot for only 1 year (2001). In addition, statistics reflect outbound mobility only to the 16 countries, and there is no information on "quality" of scientists.

The findings from the GlobSci survey indicated that patterns of mobility vary considerably across the 16 countries studied. The major reason individuals return to their home country is for family and personal reasons, not because of disparate opportunities. Additionally, mobile scientists were found to be more likely to establish international links, have links with larger numbers of countries, and exhibit superior performance on international collaborations. They were also more productive than nonmobile scientists and returnees, and results persist even after instrumenting for mobility. Finally, graduate students and postdoctoral fellows are drawn to study in the United States disproportionately because of reputation of institutions, financial support, and perceptions of how U.S.

study affects their career paths. Lifestyle issues were found to be a discouraging factor for studying in the United States.

Stephan closed by noting that surveys can take the research into international mobility of scientists and engineers only so far. She encouraged NCSES to continue to work with agencies in other countries to develop systematic ways to collect data providing consistent longitudinal information about internationally mobile scientists and engineers. More generally, she added, more thought needs to be given about how to benchmark U.S. data with data from other countries on the production and movement of human capital.

During open discussion, Rosemarie Ziedonis (Boston University) added that using individual-level data was also important for tracking the mobility of scientists and engineers, both career-wise and employment level-wise, across states and regions *within* the United States.

Jason Owen-Smith (University of Michigan) elaborated on some of Stephan's themes with a presentation on the use of university administrative data for studying the research process, specifically the role of network creation in enhancing productivity. His focus was on people as the "vectors of innovation, science, entrepreneurship" and on the structures and processes underlying scientific discovery and training in university settings. He explained that the Institute for Research on Innovation and Science (IRIS), of which he is executive director, serves as a "global source for data to support fundamental research on the results of public and private investments in discovery, innovation, and education." He noted IRIS offers useful opportunities for expanding the kinds of administrative data discussed by Stephan and others.

A standard theory of innovation and discovery is that these processes involve a combining of knowledge or techniques that have not been combined before. Advances occur when ideas come into proximity with one another, even when there is no reason a priori to expect that such a network would include these particular pieces. Every university maintains administrative data, compiled for purposes other than research that can be exploited to advance the understanding of these processes. Owen-Smith provided a visualization of how different kinds of data—whether on sponsored projects, human resources, procurement, or a host of other areas—can be combined to map a collaboration network. Constructed from UMETRICS data, nodes of individuals tied to one another—for example, being paid by the same sponsored project in the same year, as coauthors, or as members of co-patenting networks—can in principle be shown. Similarly, different parts of the network associated with the dominant topics can be mapped.

Owen-Smith asked what contributes to the growth of these idiosyncratic networks that put topics together and people together. An example of a research topic around which a network has formed is hepatitis C, which is a big problem for homeless teenagers. Clinicians and consultants work on one dimension of the problem, while specialists on liver cancer and organ transplants work on another dimension. This is an instance where social science expertise and bench science expertise is bridged by clinical expertise. With the right administrative data, Owen-Smith showed how it is possible to map the knowledge and collaborative space of a whole campus and to examine implications of the positions of individuals and teams (or even departments and programs) for scientific training and later career outcomes. This is interesting because there is dramatic campus-to-campus variation across universities, and it has implications for actionable planning policies, he said.

One question that the IRIS team has asked is how to use the expansive set of university administrative data to explore the relationships between where people are situated in physical space and how networks evolve. Using administrative directory data to position every investigator in a two-building space, they mapped out how proximate they were to one another and calculated what was called a functional zone. For every pair of individuals, the number of linear feet of overlap for those two paths was established as a naïve proxy for how likely they were to bump into each other and talk as a matter of course during their daily work.

Next, the team matched the physical space data to data on grant applications, both funded and unfunded, drawn from sponsored projects, institutional review board (IRB) applications, and animal use applications, over approximately a 15-year period for everyone in the two buildings. They also identified instances in which these activities were done jointly over a 3-year period. The research found that a 100-foot increase in functional overlap between two investigators was associated with a 14-19 percent increase in the likelihood of them forming a new collaboration (e.g., an IRB, animal use, or grant submission). Similarly, conditional on grant submission, the same 100-foot increase was associated with an 18-20 percent increase in the likelihood of a funded joint project. Such findings are indicative of how the idiosyncrasies of space and capital investments shape networks that produce findings. In addition, Owen-Smith noted, the entire approach is based on an administrative data platform that serves as a model for exposing interesting research questions that can be addressed.

During open discussion, Agarwal reiterated the value of combining data. The dream, she said, would be to integrate proprietary datasets with sources such as the LEHD and SESTAT, or IRIS. A big advantage of SESTAT is that it includes rich data on motivations and preferences, which are absent from most of the secondary data from censuses or on patents. She argued that, if the notion that both abilities and aspirations matter in determining innovation is taken seriously, then combining data that capture factors about managerial practices, as described by Gambardella, and networks of physical space, as described by Owen-Smith, is needed. From her perspective, it would represent a big step forward to move away from product-based innovation concepts toward measures that capture where the innovative activity is occurring and why it is occurring.

Stephan pointed out that some of the data improvements proposed during the session are not that expensive and that linking across sources can be an efficient way to expand information. For example, the Survey of Earned Doctorates covers practically the entire population of new Ph.D.s in the United States. It needs to be matched to other sources, such as the SDR or census data, in order to study what happens to innovative workers over time. Owen-Smith added that many costs of data linkage projects are social, not technical. They materialize in the form of negotiating memoranda of understanding, maintaining security, and managing data. Currently, he said, people do this individually, and effort is being repeated in different ways that are slightly incompatible and sometimes not well documented. One benefit of taking on these tasks as a community, he asserted, is to streamline linkages, systematize data use, and broker between data producers and parts of the federal statistical and science system to maximize the value of investments in data.

One of the greatest challenges to creative data linking is the lack of stable individual identifiers outside of the context of the Census Bureau's Protected Identification Keys (PIKs), which can only be used under very restricted conditions in Federal Statistical Research Data Centers (FSRDCs). Regulatory and other problems arise in trying to uniquely identify scientific and other academic authors and contributors in such sources as student record transcript data or workforce information. Microdata that are shared and curated by a community require development of persistent identifiers. Lucia Foster (Census Bureau) noted that her organization was in the process of developing unique identifiers for the projects taking place within the FSRDCs and that, as they bring in other outside agencies into the FSRDCs, they plan to develop datasets with unique identifiers to identify people in datasets over time.

5

Measuring Public-Sector Innovation and Social Progress

A motivation for the workshop was the belief that comprehensive measurement of the benefits generated by innovation requires capturing data on a range of activities that goes beyond those traditionally tracked, such as research and development (R&D). Innovation that takes place in the public sector is one area to be explored. Innovation that advances the public good or takes place in ways that accrue outside the market is another. Both topics have featured in discussions about revision of the *Oslo Manual* and in planning for the next Blue Sky Conference convened by the OECD.

Presenters during this session of the workshop addressed questions that included the following:

- What approaches to innovation measurement can be developed for capturing activities that take place within the public sector?
- Can metrics be developed that go beyond the contribution of innovation to efficiency, effectiveness, or quality in the production of market goods and services?
- Specifically, can the successful exploitation of ideas in the provision of public goods and in the promotion of societal well-being more broadly be measured?
- Are there negative social impacts of innovation that should be considered in understanding of the outcomes of innovation?

MEASURING PUBLIC-SECTOR INNOVATION AND SOCIAL PROGRESS AND INNOVATION INDICATORS FOR THE PUBLIC SECTOR

Fred Gault (United Nations University–Maastricht Economic and Social Research Institute on Innovation and Technology) led off the session with a discussion of the variables that factor into public-sector innovation measurement. He began by proposing an inclusive working definition of innovation:

> An *innovation* is the implementation of a new or significantly changed product (good or service) or process (production or delivery, organisation, or marketing).

And:

> A new or significantly changed product is *implemented* when it is *made available to potential users*. New or significantly changed processes are implemented when they are brought into actual use in the operation of the institutional unit, *as part of making product available to potential users*.

He pointed out that the clause "making the product available to potential users" provides a way of broadening the scope to include the public-sector domain and households.

Currently, there is no equivalent to the *Oslo Manual* or the Community Innovation Survey (CIS) for measuring innovation in the public sector, although both provide models for guidelines and measurement. One problem to be overcome is the need for registers for public-sector institutions comparable to those for businesses, he said, noting the Nordic MEPIN project has made a major contribution in this area (Bloch and Bugge, 2013).

Elsewhere, there has been experimental and evolving work including in the European Union (EU): the Innobarometer (2010)[1] attempts to measure the activity of public-sector innovation; the *Report on Public Sector Innovation* (2012)[2] includes case studies; and the European Public Sector Innovation Scoreboard EPSIS (2013) takes on the topic.[3] The 2013 scoreboard includes several measures of human resources and of the quality of public services. For the latter, this is the share of organizations and public administration with services, communication, process, or organizational innovations. All the other indicators are indirect or are contextual, which Gault asserted is one of the weaknesses of such scoreboards.

---

[1]Available: http://ec.europa.eu/public_opinion/flash/fl_305_en.pdf [November 2016].

[2]Available: http://ec.europa.eu/growth/industry/innovation/policy/public-sector/index_en.htm [November 2016].

[3]Available: http://bookshop.europa.eu/en/european-public-sector-innovation-scoreboard-2013-pbNBAZ13001/ [November 2016].

Gault noted that the OECD has also pursued work on the public sector, including the *Observatory of Public Sector Innovation*,[4] the STI e-Outlook,[5] the Blue Sky Forum 3, taking place in September 2016; and the revision of the *Oslo Manual*, which started in 2015 and is expected to be complete in 2017. The *Observatory of Public Sector Innovation* does not include definitions, but it does collect observations about public-sector innovation and, ideally, about best practices. The Science, Technology, and Industry e-Outlook provides a definition of public-sector innovation that is not too far removed from the *Oslo Manual*: "the implementation by a public-sector organisation of new or significantly improved operations or products." The OECD Blue Sky Forum, which takes place every 10 years, governs OECD thinking about measurement programs for the next decade. The revision of the *Oslo Manual* could address broader definitions of innovation that would be a first step toward providing guidance on the measurement and interpretation of innovation in the public sector and the household sector. The *Oslo Manual* could remain focused on the business sector, but, he commented, if broader definitions are not considered, other organizations may rise to the challenge.

Gault assessed the current state of measurement: Innovation in the public sector can be defined, but a consensus is needed if measurement is to be consistent and comparable. Surveys have been tested, but the results are not comparable over time or across geography. The EU and the OECD are supporting work on public-sector innovation.

Relative to the public sector overall, he said, much less thought has gone into measuring social progress, equity, inclusion, public health, education, and culture.[6] He noted the first step is to be able to detect the presence of innovation through surveys or other data collection means. The next step is to ask how innovation contributes to social progress, and how it can be measured. For example, social housing tower blocks in the UK may have qualified as an innovation, but they did not deliver on social progress. Thus, as with the business sector, if value is to be measured, there must be a follow-up survey or use of administrative data. On the policy front, public-sector actions change lives. Innovation policy can act

---

[4]Available: https://www.oecd.org/governance/observatory-public-sector-innovation.htm [November 2016].

[5]Available: https://www.oecd.org/sti/outlook/e-outlook/stipolicyprofiles/competencestoinnovate/public-sectorinnovation.htm [November 2016].

[6]Although workshop participant Daniel Sarewitz noted Chris Freeman's work with the Science Policy Research Unit of the University of Sussex on innovation and social good (see http://www.sussex.ac.uk/spru/about [August 2016]); Nelson (1977); and Lundvall (1992), which includes an understanding of the role of innovation systems for improving conditions in the world.

directly or indirectly to support innovation, but measurement is required to monitor and evaluate policy once it is implemented.

Building on Gault's discussion, Daniel Sarewitz (Arizona State University) spoke about innovation and social progress with the development of next-generation indicators in mind. He began with the observation that innovation can lead to either positive or negative outcomes, or both. Research is naturally focused on players directly involved in the innovation system, while little thought is given to those who are excluded from its benefits or who are negatively affected. The Internet, for example, sparked destructive progress in a number of ways, he said. While revolutionizing the diffusion of information, it also all but destroyed traditional journalism, for example. Financial innovations led to the subprime mortgage debacle of the late 2000s, which affected broad swaths of society. Thus, the value of innovation often depends on the perspectives of particular segments of society that are affected. He noted that the Luddites smashed power looms in the 19th century because they were going to be displaced from their jobs. They were part of an innovation episode, but they would not be a part that would typically be accounted for in measurements.

Sarewitz also urged workshop participants to keep in mind the idea that innovation and its consequences are sometimes political and normative, arguing that the presence of uncertainty and political elements means that convergence on a canonical consensus set of indicators is not possible. However, acknowledging the normative aspects of underlying processes that lead to both good and bad consequences in a complex society allows for a broader understanding of innovation. As a political process, beyond the winners and the immediate participants in the innovation system, it involves those who may not have as a conspicuous voice in that system.

INNOVATION POLICY

Charles Edquist (CIRCLE, Lund University), speaking from his perspective on the Swedish National Innovation Council, addressed three topics: (1) the conceptual basis for the design of innovation policy, (2) the usefulness of the Innovation Union Scoreboard, and (3) innovation-related public procurement.

Edquist defined innovation policy as encompassing all actions by public organizations that influence innovation processes. A less than straightforward question, he said, is "What actions should be taken by policy makers and what should be done by private organizations?" He suggested that two conditions must be fulfilled for public intervention to be warranted in a market economy. First, private actors must fail to achieve the objectives formulated—a "problem" must exist. Second, public actors must have the ability to solve or mitigate the problem. In short, policy makers must identify innovation policy problems and their causes.

In order to identify whether a country or region is doing well or badly with regard to certain kinds of innovations, the innovations need to be measured, Edquist said. Ideally, he continued, innovation indicators must be comparative between countries and regions since there is no objective optimality. The Innovation Union Scoreboard (IUS, or European Innovation) is one such attempt to measure innovation.[7] It is published every year and is intended to have an impact on policies. It includes 25 indicators that measure everything from R&D input to product innovation output. While recognizing its value in concept, Edquist identified a number of problems with the IUS, which claims to measure "EU Member States' Innovation Performance" by calculating a Summary Innovation Index (SII).

Edquist claimed the SII fails to live up to its mandate, which he said means that it provides misleading information for politicians, policy makers, researchers, and the general public. The problem is that the index calculates a simple average of the 25 individual indicators, with each factor given the same weight. Among the indicators are both inputs, such as R&D expenditures, and outputs, such as actual product innovations. No distinction is made between them, which renders the underlying method akin to taking the average between the total production and the number of employees in a firm and calling it performance. In this sense, he said, the aggregate index has no meaning.

Given this methodological problem, Edquist and Zabala-Iturriagagoitia (2015) have worked on an alternative approach based on four input indicators and eight output indicators, using only IUS data. The input indicators refer to the resources (human, material, and financial; private as well as governmental) used to create innovations, including bringing them to the market. The output indicators refer to new products and processes, new designs and community trademarks, and marketing and organizational innovations, which are either new to the market and/or new to the firm and are adopted by users. Outputs are divided by inputs in order to calculate the efficiency of innovation systems at the national level.

The results from this calculation are very different from those derived using the SII. If nothing else, this shows the inherent arbitrariness of aggregate measures based on indicator scoreboards. If summary indicators are to be calculated, the conceptual and theoretical basis has to be clear, and Edquist noted that effort must be made to understand details

---

[7]See http://ec.europa.eu/growth/industry/innovation/facts-figures/scoreboards_en [August 2016].

driving the dynamics of innovation systems. Only then can policy instruments be selected to solve or mitigate the problems, he stated, that is, if the main causes of the problems are known.

Edquist turned next to innovation policy, which he argued is far behind innovation research. The failure of the interaction between research and policy suggests the need for "a holistic innovation policy," defined as a policy that integrates all public actions that influence or may influence innovation processes in a coordinated manner. It includes actions by public organizations that unintentionally affect innovation and requires a broad view of innovation systems, including all of the determinants of the underlying processes. Edquist identified 10 important activities that are hypothetical determinants of the development and diffusion of innovations:

1. R&D
2. Education and training
3. Formation of new product markets
4. Articulation of quality requirements
5. Creation and changing organizations
6. Interactive learning
7. Creating and changing institutions
8. Incubation
9. Financing of innovation processes
10. Consultancy services

These activities also play a role in the public procurement of innovations. Edquist noted that total public procurement represents between 10 and 20 percent of gross domestic product in the European Union (around 2.3 trillion euros). This amounts to perhaps 40 to 50 times the size of the public R&D budget.

To analyze procurement demands, an assessment of what is bought in the public sector in functional terms (as opposed to products, goods, or services) is needed. In other words, policy makers should focus on the problems to be solved through innovation and spending rather than on products purchased. For addressing traffic noise in a neighborhood, for example, a local government should think in terms of buying a 1-decibel reduction in noise, not a specific item, such as a fence. The former encourages creative, competitive thinking about the effectiveness of alternatives, such as plantings, quieter road surfaces, or enforcement of slower speeds. This kind of procurement strategy is a powerful innovation policy instrument and, according to Edquist, it is starting to be used systematically by the Swedish National Innovation Council.

Kaye Husbands Fealing (Georgia Tech) continued the discussion of innovation in nonmarket contexts, specifically as it applies to public health outcomes. She focused her comments on a study on food safety and security funded by the U.S. Department of Agriculture as an example of innovation that directly affects public welfare.

The overarching question illustrated by her example, she said, is, "What are the social returns to public and private expenditures on science, technology, and innovation activities—and how can they be measured?" In the area of food safety, patents can be reviewed but, as argued by a number of participants in other contexts, this offers a highly truncated accounting of innovation. In the case of food safety sciences, new knowledge and innovations are often shared and not patented—they are not private goods. Additionally, the economic returns to federal (or other) government investments in food safety research (and other areas) cannot only be accounted for in terms of dollars and cents. Some of the value comes in the form of the development of new talent (as discussed by Stephan and others), and some will be the development of new ideas that spill over into other areas. For all of these reasons, she said, a broader measure (and indeed concept) of outputs of R&D is needed. If a goal is to have comprehensive measures of innovation inputs and outputs, particularly in the public sector, these are the kinds of factors that have to be assessed.

Husbands Fealing attached some figures to motivate the importance of food safety research. In 2011, the Centers for Disease Control and Prevention (CDC) estimated that 48 million cases of foodborne illnesses and infections led to 128,000 hospitalizations, 3,000 deaths, and $14.1 billion (in 2010 dollars) in economic burden. The CDC also estimates that one in six Americans is sickened by foodborne disease each year. She said this is a serious social problem that warrants investment in new knowledge and innovation on a regular basis, and the CDC identifies food safety as a "winnable battle" in the public health sector.[8]

Husbands Fealing noted that investment in research solves, or begins to solve, social and health problems, but few agreed-upon measures exist for assessing the value of this work. In the case of food safety, assessment involves tracking steps throughout the entire value chain from farm to table. With this goal, her research team is developing frameworks and techniques for measuring outcomes from federally funded research targeted at the agricultural sector in general and food safety in particular. The investigation has raised the following key questions:

- What expenditures have been made and how have these expenditures changed over time?

[8]See https://www.cdc.gov/winnablebattles/ [August 2016].

- Who is doing the research (principal investigators, graduate students, postdoctoral fellows, and staff scientists)?
- What kinds of jobs are taken by recent graduates trained in food safety? How do these early career activities relate to graduates' career paths? What is the role of food safety research funding in graduate student training?
- What are the outputs of federally sponsored academic research? How are the results transmitted to the scientific and private sector, commercialized, and effective for the social good?
- What is the best method for funding research leading to the most significant breakthroughs?

She said answering these research questions requires a coordinated data structure (as illustrated in Figure 5-1, which shows the emerging UMETRICS database linkages discussed earlier by Stephan). Such a structure should include project-level data (e.g., payroll records); natural language text analysis to identify food-safety research; and linkages among researchers, students, postdocs, patents, publications, dissertations, employment, and enterprises to be able to measure technology transfer.

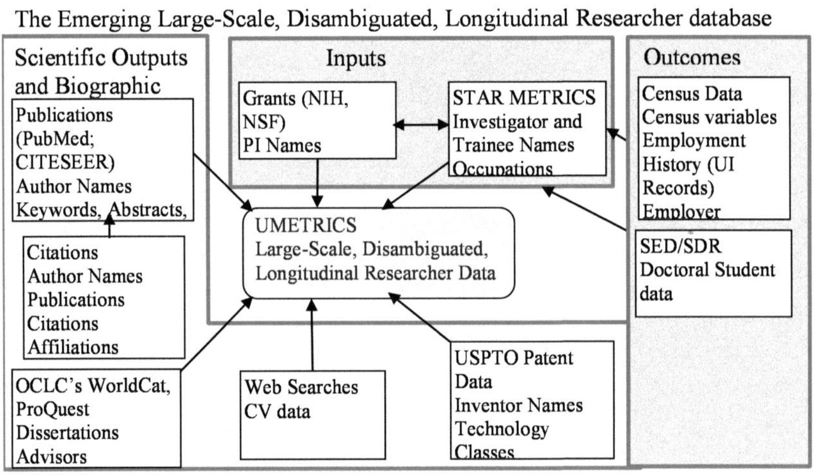

FIGURE 5-1 The emerging large-scale, disambiguated, longitudinal research database built on UMETRICS.
SOURCE: Workshop presentation by Kaye Husbands Fealing, May 19, 2016; figure from Lane et al. (2014, p. 9).

Linking the data from these sources, the team can begin to see how much money is spent on food safety research and related activities over time. This involves not only general funding by the National Science Foundation in biology, but also a number of other fields and disciplines organized in a taxonomy designed to shed light on which grants actually relate to food safety and then tracing them through the system using all the data linkages shown in the figure.

Using the food safety example, Husbands Fealing demonstrated the possibilities of measuring the role of innovation for social goods. As in environmental, health, and other research areas, she said, there is a lot of policy value to be gained from improving assessments of the return—the social benefit—from such funding on science.

HOUSEHOLD-SECTOR INNOVATION

Continuing with the theme of nonmarket production, Eric von Hippel (Massachusetts Institute of Technology) discussed household-sector innovation and made the case for why it should be measured. He noted that household-sector innovation is quantitatively very large but is not currently being measured; many activities are not even defined as innovation because the outputs are diffused for free. He also said not measuring household-sector innovation creates distortions in public policy discussions that determine what governments choose to do.

By way of background, von Hippel noted that, in economics, consumers are traditionally viewed as passive users of producer-developed products and services, and not as developers themselves (Schumpeter, 1934). In reality, however, as revealed by nationally representative time use and other kinds of surveys, an enormous amount of innovation originates from the household sector. Eliminating certain activities (e.g., those resulting in little functional novelty), von Hippel et al. (2010) found that 6.1 percent of the UK population innovated within the past 3 years, developing or improving a product for themselves (the numbers are comparable, but slightly lower for the United States). This amounts to 2.9 million people engaged in such activities, whereas in the entire United Kingdom only 22,300 R&D professionals work on developing products for consumers. The amount that households are spending is not measured at all. Inferring from the above statistics, however, it is reasonable to suspect that the level of household expenditures on innovative activities is not very different from the amount that firms are spending on R&D.[9]

---

[9]Workshop participant Jeff Oldham (Google) ran a Google Consumer Survey overnight between the days of the workshop that asked, "Did you improve or create a product or process in your free time during the past year?" Twenty percent of the more than 900 respondents

Next, von Hippel provided several examples of household-sector innovation. The Ushahidi Crisis Map service in Haiti was a very important reporting tool for providing free disaster information during the 2010 Haitian earthquake. It was developed in the household sector using the Ushahidi open-source software to post information about friends' and relatives' whereabouts.[10] In the medical sector, two people developed an artificial pancreas ahead of medical producers and diffused their design for free through the Nightscout Foundation. (This work involved creating predictive software that provides real-time recommendations for insulin and embedding the software in a closed-loop device to deliver the recommended doses.)[11] Market-based producers will follow after approval by the U.S. Food and Drug Administration (FDA).

The policy question that arises is whether these kinds of innovations can be nurtured as opposed to quelled. Is it possible, von Hippel asked, to equip individuals and households with clinical trial software so that they can engage in grassroots innovation processes themselves? Are FDA regulations set at the right levels of stringency? Just as intellectual property (IP) enhances innovation by producers, it is important to think about public support for enhancing diffusion among household-sector innovators. These, he said, are policy choices.

Turning to a possible measurement strategy, von Hippel identified two elements in the definition of a "free" innovation: (1) it is developed at private cost by individuals during their unpaid leisure time and (2) design information is unprotected by the developer and potentially acquirable by anyone "for free," representing a kind of market failure. This means that there are no transactions during the process that can be tracked by ordinary economic measures (e.g., sales or patents). Another characteristic is that less than 10 percent of these individuals try to protect their innovations, and they rarely spend much time diffusing them. As a result, IP is not a central feature of this system.

The above-described types of innovation are not measured at all in official statistics. One reason is that free innovation does not fit the *Oslo Manual* definition of an innovation, described earlier. The consequence is that a free innovation—even one diffused millions of times for free over the Internet—conforms to the definition only if it is introduced to the market. Even then, it is credited to the producer that commercialized it, not the person who developed it. As a result, estimates of the effective-

---

who replied over the 12-hour period prior to day 2 of the workshop said that they had innovated during the past year.

[10]See https://www.ushahidi.com/ [August 2016].

[11]See https://diyps.org/, and http://www.nightscoutfoundation.org/about/ [August 2016].

ness of producer R&D and the apparent importance of IP are overstated. Meanwhile, free innovation is understated, and free innovators, who are often innovation pioneers, are undercounted. The distortions created by not measuring free innovation, von Hippel argued, carry policy implications. In order to correct for this weakness in the innovation measurement system, von Hippel proposed that social surveys are needed to supplement business surveys such as the CIS.

Working with colleagues Alfonso Gambardella and Christina Raasch, von Hippel showed that free innovation increases social welfare (Gambardella et al., 2015). Because free innovation competes with producer innovation, prices may be lowered—for example, when the Linux operating system is provided for free. Free innovations are also a source of supply designs and free complements. Also, there are many things that producers make that require complements that are only provided for free, because there is no market for them. Product users—who are often early but invisible innovators—frequently play this role. Von Hippel gave the example of mountain bicycles. Mountain biking techniques are not developed in a commercial market; they are diffused peer to peer in a way that is detached from producers. Producers benefit because technique innovators increase the value of mountain bikes and may increase sales. Von Hippel and his colleagues also studied whitewater kayaking and discovered that, over 50 years, among 100 of the most important innovations, 80 percent were introduced by the users of kayaks and picked up for free by the producers (Hienerth et al., 2012). If producers had had to develop all these designs, their budgets for R&D would have needed to be three times larger. These processes take place throughout the economy—this kind of feedback is common with scientific and medical instruments, for example.

Von Hippel concluded by offering some guidance to NCSES, other statistical agencies, and OECD. He urged that the *Oslo Manual* include nonmarket diffusion. Its exclusion from the definition is a barrier to recognizing that activities taking place beyond the market can be innovations. This is the gateway problem, he said. Myopic focus on the market is an outdated way of thinking, reflecting a time when the market was the dominant mechanism for diffusing things. In an Internet-based world, free diffusion is prominent and should be reflected in new measurement guidelines.

In open discussion, Ben Martin said he largely agreed with this view, but he added qualifications about the intrinsic limits of indicators. He agreed that the scope of indicators should be extended to cover a wider range of innovations, but he expressed concern about Gault's proposal to change the wording in the *Oslo Manual* definition from "improved" to "changed." He cautioned that, whether in the market or nonmarket context, there could be a danger of bringing in incremental product changes that do not have an impact on economic or social outcomes and well-being.

Gault responded that measurement is something that should be pursued with as few normative terms as possible. If a new financial product based on subprime mortgages has been produced and put on the market, for example, innovation measures should capture this. That the product is then diffused and goes on to harm the economy is exposed only after the fact through normative assessment. It is not so clear that the assessment itself should be part of the innovation indicator system, he said, as opposed to being the subject of subsequent research measuring the outcomes associated with the innovation.

6

Regional Innovation Models and Data Needs

Introducing the workshop session on regional innovation models and data, Maryann Feldman (University of North Carolina) observed that the locus of innovative activity is inherently subnational—certainly substate and, in many cases, very local—and many spillovers and complementarities of innovative activity take place at the local level as well. Yet data collection is focused on aggregate, often national levels of geography. This workshop session was intended to provide guidance for addressing this disconnect and to highlight important work developing regional data infrastructures. Much of the focus was on entrepreneurship, one dimension of innovation.

REGIONAL MEASURES OF INNOVATION AND ENTREPRENEURSHIP

Catherine Fazio (Massachusetts Institute of Technology) began the session presenting estimates of the quantity and quality of entrepreneurship at localized levels. She began with the observation, based on research (e.g., Decker et al., 2015; Haltiwanger et al., 2014), that the benefit of startups for economic growth spring from a handful of fast-growing, young companies such that the distribution of the sources of job and wage growth is highly skewed. The policy challenge, as expressed by Robert Litan (2010), is to propel more startups that are capable of meteoric growth. The problem is the difficulty to predict at the time of founding which firms are likely to survive and grow. For example, in 1994, what distinguished a startup bookstore that would remain unchanged for years (or go out of business) from Amazon? At the time of founding, it is typically difficult to discern any difference based on traditional metrics. The measurement challenge is how to identify and map growth potential from the start.

Fazio said traditional measures provide important information. For example, quantity-based measures from the Census Bureau's Business Dynamics Statistics (BDS) document a 30-year decline in business dynamism. Figure 6-1 indicates a significant decline in the rate of firm entry, providing one measure that the U.S. economy has become less entrepreneurial over time.

In contrast, Figure 6-2 provides an indication that high-growth entrepreneurship is on the rise—so much so that there has been concern over whether or not there is a bubble and, in some places, even whether the bubble has now burst (Rosoff et al., 2015).

The above metrics lead to conflicting implications, Fazio explained. Measures that account for firms' initial heterogeneity and growth ambi-

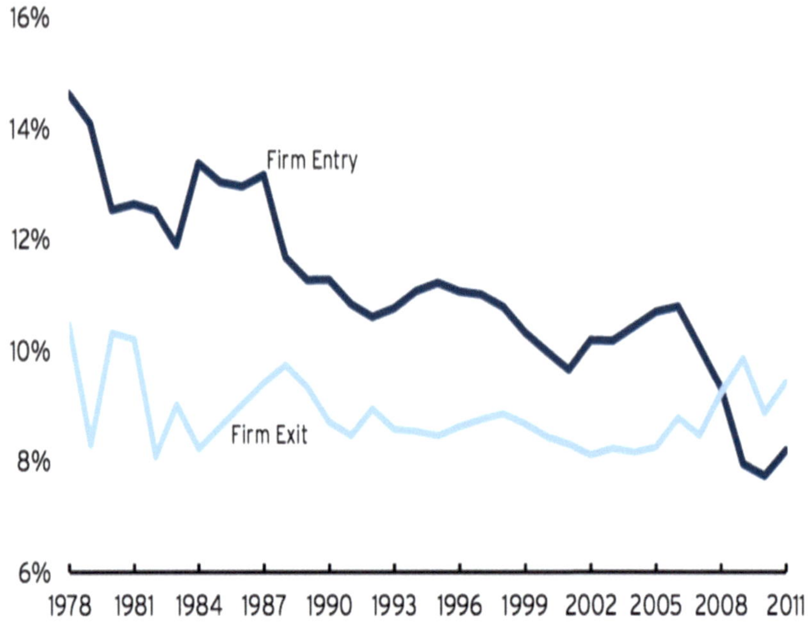

FIGURE 6-1 Rates of U.S. firm entry and exit, 1978-2011.
SOURCE: Workshop presentation by Catherine Fazio, May 20, 2016; figure from Hathaway and Litan (2014, figure 3), based on Business Dynamics Statistics data.

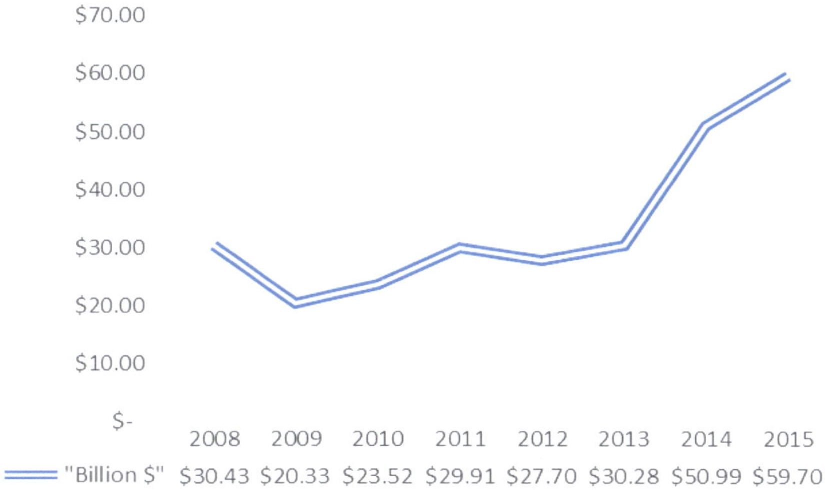

FIGURE 6-2 U.S. venture capital investments ($B), 2008-2015.
SOURCE: Workshop presentation by Catherine Fazio, May 20, 2016; data from The MoneyTree™ Report by PricewaterhouseCoopers and the National Venture Capital Association, based on data from Thomson Reuters. See https://www.pwcmoneytree.com/HistoricTrends/CustomQueryHistoricTrend [November 2016].

tion and potential are needed to fill the missing part of the story. To maximize the relevance and utility of data collection programs, the National Center for Science and Engineering Statistics (NCSES), OECD, and statistical agencies need to recognize that small to medium-sized enterprises (SMEs) and innovation-driven enterprises (IDEs) are different. To capture these differences, information systems should be broadened to include measures of entrepreneurial quality that enable policy makers to evaluate SMEs and IDEs at the time of formation.

Fazio proceeded to outline new approaches for measuring entrepreneurial quantity and quality and to describe findings emerging from research based on these measures. The quality measures attempt to estimate underlying growth potential of startups, drawing on characteristics observable at or near the time of founding, and deriving conditions that yield consistent population-level statistics. They include the Entrepreneurial Quality Index, designed to indicate the average growth potential of any given group of new firms; the Regional Entrepreneurship Cohort Potential Index, the number of startups within a region that relates to expected growth outcomes; and the Regional Entrepreneurship Acceleration Index, which is the ability of a region to convert entrepreneurial potential into realized growth (see Guzman and Stern, 2016). These measures generate novel characterizations of entrepreneurial ecosystems over time at different levels of geographic granularity, she said.

The approach described by Fazio builds on three interrelated insights to develop a new data source on innovation-driven entrepreneurship. The first is that business registration is a practical requirement for growth. Business registration represents, at the state level, a robust sample of entrepreneurs who are at similar foundational stages of their process. Business registers are comparable over time and place, but they do not include all data that are needed. They do not include nonregistered firms, sole proprietorships, unincorporated or self-employed people, or the household innovators described by von Hippel.

The second insight is that markers of entrepreneurial quality are observable at or near the time of business registration. Firms with the potential and ambition for meaningful growth outcomes often have different "startup characteristics," some of which are directly observable within business registration records. Firms often aim for a certain level of growth and then strive to achieve it. For example, their growth ambitions are sometimes reflected in the names that they choose. The Corner Café has a different growth ambition than Akamai Technologies or Amazon, Fazio asserted. Patent filings made by early-stage ventures are also directly observable and may signal a new firm's potential and ambition for growth. These kinds of characteristics captured by various data sources can be distilled into measures and then scored depending on whether those characteristics were exhibited or not.

The third insight highlighted by Fazio is that meaningful growth outcomes can be observed with a lag, creating the potential to map observed growth with characteristics. Rather than assuming a relationship, the relative importance of different factors can be investigated by developing a predictive model of growth or entrepreneurial quality based on startup characteristics. Among the findings from such a model (Fazio et al., 2016, Table 1): A startup that is a corporation is 400 percent more likely to grow compared to one that is not, and a startup named after a founder is 70 percent less likely to grow compared with one that is not. If a startup in the United States exhibits a combination of positively correlated characteristics—such as having patents *and* a registration in the state of Delaware—the probability of growth jumps astronomically. Such findings can be used to construct the underlying probability of growth at the time of founding for every registered firm.

Fazio described these startup characteristics as digital signatures of growth ambition and potential, not causal drivers. In a test of skewness among new firms using Estimated Entrepreneurial Quality, Fazio et al.

(2016, endnote 38) found that 65-72 percent of realized growth events fall within the top 5 percent of the model's estimated entrepreneurial quality distribution, and 51 percent of realized growth events fall within the top 1 percent of the distribution of startups. In other words, growth potential is a key dimension of heterogeneity among newly founded firms.

Fazio turned to results from Guzman and Stern (2016) based on regional entrepreneurship cohort quality indexes (which can be used to estimate the expected number of growth events) and the ecosystem acceleration index (which indicates the ratio of realized to expected growth events in a region). These tools can be used to map where firms with high growth potential are being founded and located. In a Boston area analysis, for example, the Cambridge Innovation Center and the cluster of biotech lab spaces clustered along the Charles River appear prominently. When the above described quality-based measures of innovation-driven entrepreneurship are added to quantity measures, areas such as Kendall Square become more prominent, while areas with standard-type businesses become less so.

As a policy tool at the subnational level, these models enable evaluation of innovation-driven entrepreneurship at a more granular level than has been possible, Fazio observed. They permit more tailored analyses and could contribute to the development of more targeted policy interventions that can be tested through experiments. For regional policy makers, these analyses can help identify whether IDEs are being started, which firms have high potential, what that potential for growth is, and which sectors they are in (including potential opportunities for cluster-based IDE). Trends can be identified, such as where firms are locating, whether they are moving, if realized growth is matching expected growth, if the ecosystem is conducive to growth, and if there is coincidence with any government programs. Finally, she said, these analyses enable diagnosis of challenges and opportunities with respect to innovation-driven enterprises at a granular level.

Fazio concluded by identifying opportunities for collaboration among researchers, NCSES, and other statistical agencies to advance this research program. She suggested priority areas including: developing quarterly metrics of entrepreneurial quality as a regularly produced statistic for the United States; connecting entrepreneurial quality with alternative measures of performance via the Census Bureau's Longitudinal Business Database (LBD) microdata; and extending the evidence base for innovation and entrepreneurship program evaluation.

More generally, Fazio concluded it would be possible to implement predictive analytics in statistical approaches to complement other data being collected on innovation—for example, to assess the entrepreneurial potential of patent-holding entities and to tailor entrepreneurial quality measures to complement *Science and Engineering Indicators* with measures of local science-based entrepreneurship.

Following up on several of Fazio's themes, Sheryl Winston Smith (Temple University) discussed microlevel foundations for measuring innovation and entrepreneurship using novel data created by financing programs on founders, startups, and early hiring characteristics. She shared insights about the changing face of regional innovation—who is involved, what types of businesses enter, where they are entering, and how they get launched—derived from research on seed accelerators. Early-stage financing has particular implications for who enters into science, technology, engineering, and mathematics (STEM) entrepreneurship and for the trajectories of these new ventures, she said.

A key question raised by Winston Smith is whether current measures are capturing modern entrepreneurial processes and new aspects of the ecosystem. If not, she asked, what types of novel data can be collected and analyzed in order to gain deeper insights? She discussed both the opportunities associated with new kinds of data—for example, sites that provide profiles of people, companies, or both, such as LinkedIn and CrunchBase—and caveats and cautions.

In the search for answers to the kinds of questions highlighted by Winston Smith, researchers can take advantage of the fact that launching ventures typically requires seed capital. Entrepreneurs with high-growth potential ventures typically reach a stage where they need formal outside equity financing. The traditional next step has been angel capital, particularly professional angel groups, which are regionally distributed.

Seed accelerators provide a distinctly different model. They operate on a short, finite time period, typically about 3 months. People apply to enter the accelerator events, and those who are accepted proceed as a cohort through intensive boot camp periods. The process culminates in pitch events—"demo days"—where founding teams pitch their proposals to a roomful of investors and potential acquirers. There are strong elements of mentorship and peer learning, but competition as well. Winston Smith cited a number of prominent companies that have gone through startup accelerators. Dropbox originated in a seed accelerator held by Y Combinator in Boston in 2007; Airbnb went through Y Combinator in Silicon Valley in 2009; and BB-8 started life with a company called Orbotix, now Sphero, that went through a Techstars seed accelerator program in Boulder in 2010.

Winston Smith cited several figures to provide a sense of the magnitude of the seed accelerator trend (Loeb, 2014). Companies that have gone through Techstars average about $1.6 million in outside venture capital financing after leaving. The average valuation of Techstars' alumni is $4.3 million; the total value of the companies is more than $1.5 billion.

The total valuation of Y Combinator alumni companies is more than $65 billion. Eight Y Combinator companies are valued at more than $1 billion each. By comparison, angel-backed companies—the traditional source for early-stage financing—have a median premoney valuation of $3 million. For 2014, 870 angel deals involved about $1.65 billion in total rounds of financing (Angel Resource Institute, 2014). These figures indicate that accelerators are a phenomenon (or an innovation) on par with the impact that angel groups have traditionally held in the economy.

For measuring entrepreneurial activity, including the financing dimension, the issue of a skewed distribution rises to the fore once again. It is known ex ante that most firms—even most high-growth potential firms—that launch will fail. There are a few potentially very big successes and also a large middle distribution (Guzman and Stern, 2015). Issues also arise concerning typical milestones that are very rare events. An initial public offering (IPO) is a rare event; even venture capital (VC) funding is rare (about 4% of firms receive VC investments at some point). It is not the characteristic path of most new ventures, even among high-growth new ventures, she observed. In order to address policy questions, the full distribution of these phenomena needs to be understood. Intangibles—learning, competition, follow-on networks, and mentorship—may matter even more than the amount of financing. These things are hard to capture in terms of totals and averages.

Winston Smith addressed how incentives and institutional structures associated with different kinds of financing affect the growth trajectory of new ventures. She noted there may be regional implications for the entrepreneurial ecosystem, including short- and long-term impacts in terms of network and syndication ties, job creations, the recycling of founders' successful and unsuccessful new startups, new investors, and new mentors. For startups, there are questions about people—the source of human capital behind these ventures. What prior experience, networks, background, and education (particularly STEM) do founders and hires bring to the table, she queried. What are the team's funding and exit options (e.g., quitting or acquisition)? For some of these issues, it is particularly hard to get data from traditional sources. Information on failures and stagnation has to be tracked down because the stream of data typically ends along with the venture.

The accelerator process makes it possible to develop a novel microdata set to bear on some of these key questions. Winston Smith's team has been able to assemble data on 25 cohorts from Y Combinator and Techstars (the two most established accelerators) covering the 2005–2011 period. They have been able to track every outcome associated with 394 startups, 933 founders, and more than 15,000 hires through 2016. The "census" of ventures that went through these two accelerator firms is diverse, both geographically and by industry. The researchers also created a matched angel sample for a similar range of industries and geographic locations for the same time period.

One finding from the team's comparison of the accelerator and angel data is that, while the accelerator events are located in a limited number of locations throughout the United States, the startups that apply to and are accepted into these programs come from a much broader set of locations. Hires from the successful startups are drawn from an even more widespread, global distribution. This means that the accelerator process is having an impact on the distribution of human capital in a large and powerful way, she said.

Winston Smith explained that data underlying the analysis include both established and novel methods (Web scraping) and hand collection. Sources such as CrunchBase, LinkedIn, CB Insights, and technology blogs were triangulated to trace the trajectory of startups from inception, to seeding by round, to various outcomes. For each startup and founding team, Winston Smith and her collaborators tracked outcomes on quits, acquisitions, VC follow-on funding, hiring (e.g., first hiring choices, timing, generalist versus specialist), and long-term growth.

Among the many findings emerging from the analysis is that entrepreneurs with STEM backgrounds make up about one-third of founding team members, another one-third have computer coding backgrounds, and about 54 percent have business backgrounds. Winston Smith et al. (2015) find that graduates from computer science programs have a particularly strong impact; even students who study other fields but go to universities that have strong computer science programs are much more likely to enter an accelerator event. Additionally, the researchers were interested in cohort heterogeneity. The distance between the founding team from the rest of the cohort can impact the trajectory of the different startups.

In terms of next steps, Winston Smith suggested that researchers would benefit from being able to generalize her team's method to a broader group of accelerators since there is not a single model for all types of founders and startups. Increasing data coverage would allow researchers to scale up the analysis and compare findings based on established sources such as census data or the Kauffman Firm Survey.[1] A set of best practices for this kind of research should also be developed, she said.

---

[1]The Ewing Marion Kauffman Foundation sponsored the Kauffman Firm Survey, which collected data from more than 4,900 businesses established in 2004, following them through 2011 (see Robb and Farhat, 2013). Kauffman is currently helping to support an Annual Survey of Entrepreneurs (ASE), conducted by the U.S. Census Bureau beginning in 2015 for 2014 data and with sufficient funding to collect data for a total of 3 years (see https://www.census.gov/programs-surveys/ase/about.html [August 2016]).

LESSONS FROM ADMINISTRATIVE AND OTHER DATA SOURCES ABOUT REGIONAL INNOVATION AND ENTREPRENEURSHIP

Rosemarie Ziedonis (Boston University) addressed the question about what can be learned from subnational, state, regional, and local government programs directed toward supporting innovative activity. Her focus was on distinguishing what she called "young gazelles in the waiting" from the majority of companies that fail along the way. It is difficult to predict which way a firm will go, but Ziedonis asserted that opportunities exist to analyze nascent companies using data, primarily administrative, that are already collected but not often easy to access.

Using the Michigan Economic Development Corporation's loan programs as a case study, she described ongoing work by Zhao and Ziedonis (2012) that leverages data on a sample of 297 proposals from 241 startups seeking funding during the period 2002-2008. A typical applicant was a 4-year-old life science company. The idea behind this research, she said, is to test the effect of public R&D financing on recipient startups in terms of a range of outcomes: their survival (based on state business registry data), their ability to secure follow-on financing from venture capitalists and federal small business innovation research [SBIR] grants, their broader business activity (proxied by information gathered from news articles and elsewhere), and their production of patents. Preliminary findings suggest that among close-call applicants—those with similar merit scores, some of which lost out on loans by a small margin and some of which won by a small margin—those that were awarded financing experienced an increase of about 20 percent in the likelihood of survival 6 years following the competition and a higher likelihood of follow-on financing from venture capitalists and greater business expansion. This finding holds for the youngest companies and those that did not have prior SBIR funding and for those that were farther away from state hubs of entrepreneurial activity. No discernible impacts were found regarding the impact of receiving funding on patent-based outcome measures.

The method that she and Zhao employed allows for treatment effects to be estimated—not perfectly, but better than in most other options. The alternatives typically involve surveying successful recipients of these awards, which indeed can be useful. But, without the comparison group, it is difficult to make inferences about what would have happened to companies had they not received funding. This follow-up survey approach has been more commonly used with proprietary, applicant-level data for evaluating the effects of public R&D grants on postdocs, on the career paths of scientists, and, in several studies, on firm-level outcomes.

Thus, while their research illustrates a useful method for program evaluation, she stressed that they had atypical access to data—including scores from a competition-based program—for an entire pool of loan applicants as well as outcome metrics for these young and small companies. Without access to Michigan government archives, the project would not have been possible. The challenge to scaling up this kind of effort is that it would require creating a clearinghouse wherein state economic development officials could, with confidence about confidentiality protection, deposit data that could be accessed by researchers. This kind of research also requires hard work on other aspects of data collection, she said. They compiled press releases and news articles capturing information about presentations, technical conferences, sales, contracts, and product development announcements.

Administrative data are extremely useful for measuring outcomes and for designing program evaluation studies. From a policy perspective, experiments of this sort reveal evidence about the marginal impact of funding—specifically, if financing allows new firms to remain in business. The approach facilitates analyses of the "stay rates" of entrepreneurial firms and consideration of which policy levers are most effective for promoting economic development. But, until this kind of data infrastructure becomes more broadly based, a number of important questions will go unanswered: Do the findings hold when smaller award sizes are involved? Do results differ for loan versus subsidy programs? And, do the results translate to other state and local environments? At this point, Ziedonis said, researchers have produced illuminating case studies, but it is hard to know how generalizable the results are.

During open discussion, Ben Martin praised the "handcrafted indicators" described by Fazio, Winston Smith, and Ziedonis. Returning to the dark-matter analogy—the aspects of innovation that are not well measured—he concluded that new lampposts were being built through this cutting-edge research. He added, however, that much of the work focuses on entrepreneurial activity and not broader aspects of innovation, and he asked about the relationship between the two. Owen-Smith responded that because much of what had been discussed involved STEM-based entrepreneurship, many of the outputs translate into innovations. Much of what these new firms produce involved traditional innovations—new products, new services, new technological products, or new applications that come to market. A lot of the relevance, he argued, is in understanding the entrepreneurial activity, and the next step would be to trace that to innovation outcomes.

Javier Miranda referred to the access issue. He acknowledged that access to census data and local datasets can be difficult, but he said the Census Bureau has in place a process for researcher use of data. Specifically, there are 23 Federal Statistical Research Data Center (FSRDC) locations created in cooperation with universities, nonprofit research institutions, and government agencies. Also, the Census Bureau has a mandate to expand its data holdings beyond federal data, and the Evidence-Based Policymaking Commission (created through the Evidence-Based Policymaking Commission Act of 2016, known as the Ryan-Murray bill)[2] is charged to consider what would be involved in establishing a clearinghouse for federal, state, and local administrative data to inform federal policies. He recognized that these are broad umbrellas, but there is the capacity to create leverage to begin coordinating data more systematically and in a way that researchers can have access.

Maryann Feldman presented research on regional innovation systems motivated by questions about the ways in which places change and evolve over time. She cited Durham, North Carolina, as one among thousands of possible case studies. There are many entrepreneurial firms located there, as well as accelerators and incubators. A major goal of recent research by Feldman and her team is to operationalize the entrepreneurial ecosystem,[3] and to study it as a temporal spatial process.

Feldman noted the literature on regional economies is dominated by two theoretically distinct concepts: the *industrial cluster* and the *regional innovation system.* Industrial clusters are defined as "a concentration of inter-dependent firms within the same or adjacent industrial sectors in a small geographic area," while innovation systems are "interacting knowledge generation and exploitation subsystems linked to global, national and other regional systems" (Asheim and Coenen, 2005, p. 1174). Cluster analysis is firm-centric and delineates how different types of firms interact with one another within a given industry and geographic space (Lowe, 2009; Porter, 1990; Renski et al., 2007).

In this literature, processes of change tend to be underexamined, she said. What is missing is the institutional richness of the systems of innovation and an understanding of all of the supports that touch entrepreneurial firms. To address this gap, Feldman and colleagues have engaged in in-depth data collection centered around Research Triangle Park, a large industrial park in between the campuses of the University of North Carolina, Duke University, and North Carolina State University. The site was very consciously created by public policy over a 60-year time period and is now defined by a configuration of firms that has evolved over time.

Research Triangle Park, which now consists of around 5,000 firms, was established near the cities of Durham and Raleigh. It has been filled

---

[2]See https://www.congress.gov/bill/114th-congress/house-bill/1831/text/pl [November 2016].

[3]A definition of the entrepreneurial ecosystem is "a set of interconnected entrepreneurial actors, . . institutions, . . entrepreneurial organizations . . . and entrepreneurial processes which formally and informally coalesce to connect, mediate and govern the performance within the local entrepreneurial environment" (Mason and Brown, 2014).

with a large number of startup firms. Even though it is a vibrant economy, the area does not make many lists of the top fast-growing regions in the country because most of them are compiled by metropolitan statistical areas (MSAs). Durham-Chapel Hill and Raleigh are two separate MSAs; if these MSAs were combined, the area's population is about the same size as that of Austin, Texas.

Feldman, Nichola Lowe, and their research team have built a relational database (represented in Figure 6-3) in order to organize as much detail as possible about firms in the Park—the year they were established, their technology, annual events (e.g., merger, funding received, closure), job counts, and institutional supports. The cumulative effect of what makes a firm successful is the multiplicity of all of these factors and interventions by the many people who have touched them and influenced their behavior. The team triangulates data sources to establish

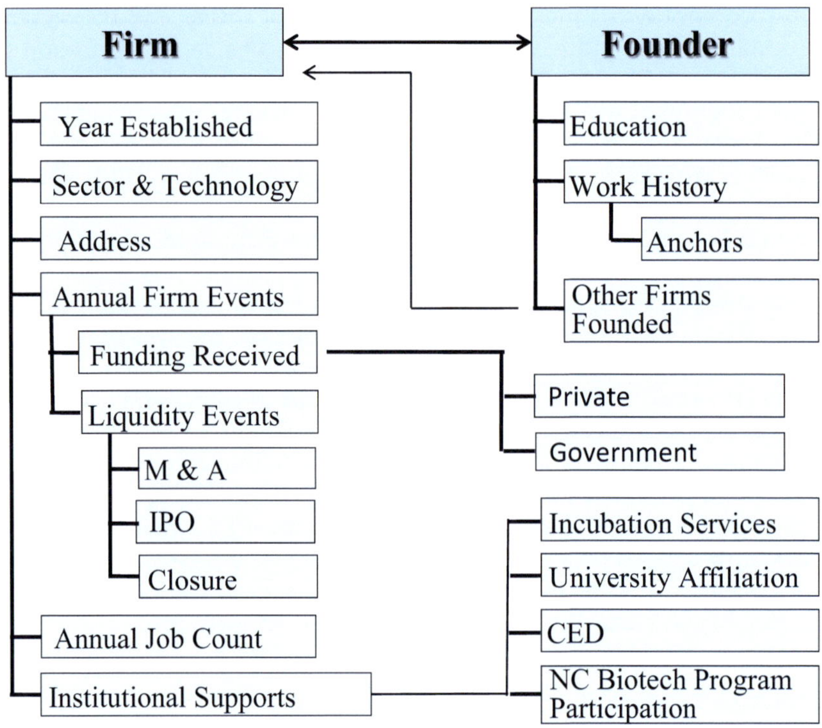

FIGURE 6-3 Research Triangle Park database of firms: Information ("firm forensics") from more than 20 sources.
SOURCE: Workshop presentation by Maryann Feldman, May 20, 2016.

when a firm begins, its growth trend, and its innovative activity. State registries are used to identify firm births and to establish starting dates. Firm growth can be tracked in terms of annual employment, sales, and revenue, numbers.

Innovative activity is tracked using data on patents, U.S. Food and Drug Administration trials, and new product announcements using web-scraping methods, some of which has been automated by the team. For every firm, the researchers look at events such as funding raised (public and private), mergers and acquisitions (M&A), initial public offerings (IPOs), and participation in incubators and entrepreneurial support organization programs. To these firm-level data, geocoded addresses can be added to examine the development and growth of the region over time and to identify micro geographies where activity has developed.

Feldman pointed out that, early on, Research Triangle Park was not very interested in small entrepreneurial firms because they are unreliable tenants. Now, however, the Park is full of technology-intensive industries that include biotech firms spawned from corporations, many of which are traced back to two prominent pharmaceutical firms in the region, GlaxoSmithKline (GSK) and Burroughs Wellcome. These large companies have spun off more new firms than the adjacent universities. About 146 first-generation spawns alone came out of GSK.[4] This in-depth analysis reveals a lot about how what large firms are doing affects entrepreneurial activity. For example, the data indicated that layoffs followed every M&A.

Feldman closed with several summary observations:

- Regional data are at present inadequate—there is not enough granularity; there are 2- to 3-year lags in publishing; and there are not enough data on the outcomes. For these reasons, her research projects have required hand-collected data.
- A new age of cyber-enabled research has begun. In addition to the growing richness of data housed at the Census Bureau's FSRDCs, Feldman noted that the amount and detail of information that people will reveal about themselves on the web is amazing. And sometimes records can be matched across sources that create even greater richness of detail. New visualization techniques are also adding value to the data being generated and linked.
- State and local administrative records need to be better exploited. Feldman suggested that an equivalent of the Evidence-Based Policymaking Commission Act of 2016 for the state level would offer great research promise. At present, little is known about how much money is spent by state and local agencies on R&D pro-

[4]See DeBruyn (2015, image showing data compiled by Feldman and Lowe).

grams. For purposes of accountability, making such data available is becoming increasingly important.

During open discussion, Wesley Cohen reiterated points made by Winston Smith, Ziedonis, and Feldman about the importance of being able to track companies that fail and what happens to the entrepreneurs associated with them. He noted that many founders of biotech firms that fail go on to benefit from the experience in ways that aid future ventures or that help them move into other areas. A data infrastructure allowing individuals involved in these ventures to be followed would no doubt produce valuable insights. Feldman reported that, of the 5,000 firms in their study, about 40 percent exited through one mechanism or another. They are easy to track when they are involved in M&A. When firms are born, there are a lot of announcements and other signals; however, she said, deaths are typically quiet. One way to tell if a firm has gone out of business is if its business registration is not renewed. In her survey, individuals involved with disappearing firms are not captured when they leave the profession or the region. Data would have to be collected systematically on a more national scale to capture the career paths of everyone involved in Research Triangle Park ventures. Cohen agreed that being able to track where talented, often high-skilled and high-powered individuals go after firm death or other events would be highly important and interesting. Feldman noted that for some kinds of studies, census data such as the Longitudinal Employer-Household Dynamics Program provide a good platform. Owen-Smith added that their UMETRICS data allow for founders involved in failures to be tracked because the tool is organized around individuals and not locations. This is the case because the data usually include people's entire career histories.

INNOVATION DATA AND ANALYSIS TO INFORM REGIONAL POLICY

Thomas Guevara (Economic Development Administration [EDA], U.S. Department of Commerce) said he spoke as a policy maker and consumer of information produced by the statistical agencies. His interest is in seeing boundaries pushed in terms of the capacity of data for informing and affecting policy and ultimately serving communities. He said his overarching premise is that the purpose of any government agency is to improve the quality of life for the citizens it serves.

When EDA started out as a public works agency 50 years ago, the perceived route to prosperity was to build more infrastructure. Implicit in this view is that prosperity is based on consumption. Guevara observed that researchers and policy makers are coming to realize that continuous growth and prosperity is not going to be achieved simply by consuming. In that spirit, his agency adopted the following definition of economic development, drawing on Feldman et al. (2016, p. 10):

> Economic development is defined as the expansion of capacities that contribute to the advancement of society through the realization of individual, firm and community potential. Economic development creates the conditions for growth and improved quality of life through innovation, lowered transaction costs, and the utilization of capabilities to responsibly produce and sell valuable goods and services.

The emphasis in this definition, he noted, is on human activity, behavior, and the way people work with one another with the resources they are given. With this in mind, EDA seeks to help communities understand their assets as well as their strengths and weaknesses, and the opportunities and threats that influence prospects for prosperity and improvement in quality of life. The agency works toward this goal by investing with local partners, he explained. It does not invest in businesses directly; rather, it provides grants to government entities and nonprofits, including universities, to take advantage of opportunities in local ecosystems. At the same time, the agency requires that businesses benefit directly from assets that they build and that they are maintained in the public trust.

Fundamentally, he said, EDA is about increasing capacity through the sponsorship of innovation-inducing activities and resources that are inputs to economic development. Guevara injected the view that underlying the innovation and development process, agency matters in terms of the ability of people to work together with scarce resources. And, he added, if agency—whether in the context of exchanging knowledge, discrete objects, or technology—matters, then complexity must be built into the measurement framework. This is the reason, in Guevara's assessment, why the research reported on at the workshop by Feldman, Fazio, Ziedonis, and Winston Smith is essential, because it involves data that provide insights into how people work together. This, Guevara concluded, is the boundary that needs to be pushed if agencies such as EDA are to become effective at investing resources to achieve outcomes that result in prosperity and improved quality of life.

Guevara raised three questions related to improving the data infrastructure:

1. What is known about agency's role in inducing innovation?
2. Can research and data collection methods be designed that ultimately lead to more precise "diagnosis and treatment" interventions for public investment, and better policy as a result?
3. Can the principles that currently define complex adaptive systems lead to greater insights about how innovation emerges?

Much of what is known about the first question can be found in the research on complex adaptive systems (e.g., Holland, 2006). The second question addresses Guevara's point about the absence of a systematic approach to data collection, analysis, and evidence to support policy and economic development. This relates directly to how the social science community does research: Researchers try things out, test and measure, and then hope to inform policy through a sense of what is learned from research findings. Learning results with some set of probabilities about what might be more likely to work and what might be less likely to work. The result is, hopefully, better policy, using health care as a great example of an ecosystem operating in a complex, adaptive system. Regarding the third question, Guevara asked whether the data being created by government agencies and in the private sector can provide the foundation for understanding innovation and its role alongside other factors in contributing to economic growth and improved well-being. He observed that some developments emerging from other disciplines—such as cluster mapping tools—may help to create new frameworks for evaluating measures of innovation and what people do to induce innovation in their environments.

During open discussion, Ziedonis pointed to growing evidence that direct intervention programs (such as those discussed by Guevara) also serve an informational role to the private capital market. While there is not a lot of consensus about why and how, research by Josh Lerner and others (e.g., Lerner and Kovner, 2015) has looked at various programs and attributed some impact on outcomes to certification whereby, when awards are won, a promising startup is declared in a sense. More research is needed, she added, about how this mechanism works. She questioned whether this is indicative of where money is being put to prototype and further develop things to a stage in which it becomes easier to communicate and actually attract private financing.

7

Innovation Measurement Agendas of the Future

A motivation for the workshop was the need for new measures of science and technology describing inputs and outcomes associated with innovative activity and, concomitantly, the need to call upon multiple modes of data. Commercial data, some recoverable from web scraping and other computer science methods, may shape future measurement in a range of areas—for example, in tracking new product introductions, quality change, prices and productivity, and product diffusion—where large datasets provide advantages in terms of granularity, timeliness, geographic specificity, and other factors. During this session, presenters continued discussing several of the data-related questions raised at various points during the workshop: To what extent can the digital revolution transform metrics in the area of innovation measurement? What are the roles of specialized surveys in innovative data collection? And, how can administrative records best be exploited?

THE CHARACTER AND DISTRIBUTION OF INNOVATION ACTIVITIES AND OUTCOMES

Scott Stern (Massachusetts Institute of Technology [MIT]) presented his perspectives on developing metrics for capturing the characteristics of innovation application and diffusion. In the process, he made the case that distinguishing between innovation that is cumulative versus that which is one-off in nature is essential to understanding the relationship between inputs to, and the skewed distribution of outputs and outcomes from, the process.

Stern opened with an illustrative example—CRISPR, a gene-editing tool that was developed with public funding in the 1980s. It was not until 2012, however—when a group of researchers at Berkeley, followed shortly thereafter by others at MIT and Harvard, figured out that the tool could be repurposed slightly—that CRISPR became arguably the single largest advance in life sciences gene editing ever. The cumulative number of forward citations of articles by these research teams on the topic is now up to about 3,500 on an annual basis (see Ledford, 2016).

CRISPR is essentially a cut-and-paste editing tool for genes that can be used by large pharmaceutical companies, biotech firms, and startups. For example, very recently, the tool was decisive in the creation of a platform for detecting Zika virus. The emergence of CRISPR is an example of the process from basic discovery to traditional science and technology to unanticipated applications—in this case, all within a 3-year period. Stern said the episode illustrates three important points: (1) innovation is inherently cumulative, (2) innovation is inherently uncertain, and (3) innovation is highly skewed in the distribution of application and in its impact across multiple dimensions.

He noted one of the key questions of the workshop was how fundamental characteristics of innovation are reflected in a measurement framework and in systematic statistical programs for collecting information about innovation in order to move beyond examples. The cumulativeness of innovation—that is, the ability to draw upon an ever-wider body of scientific and technical knowledge—is widely regarded as a critical component for idea-driven, long-term economic growth (Aghion and Howitt, 1998; Dasgupta and David, 1994; Mokyr, 2004; Romer, 1994; Rosenberg, 1982). But, Stern asked, how can it be known that innovation is cumulative, how and whether the degree of cumulativeness varies by time and place, and how it can be measured?

To test cumulative impacts, Furman and Stern (2011) examined biological resource centers, wherein biological materials used in research are deposited and made publicly accessible to future generations of researchers. It allows them to avoid having to reinvent the wheel, Stern noted. The authors identified an approach using citations data to track the diffusion of knowledge from scientific papers placed into an institutional environment that promotes cumulativeness—a biological resource center—compared with similar articles that remained in a less open system. The articles were tracked to estimate cumulative impact, measured by the rate of citations (which he noted is not the same as innovation, but an indicator of influence). By this metric, for articles placed in the biological resource centers, the authors found a more than doubling of the impact on subsequent productivity of publicly funded knowledge.

For purposes of systematic measurement, the question that arises is how to aggregate this kind of study of cumulativeness (and the role of institutions and policy in shaping cumulativeness and outcomes) beyond the level of individual "pieces" of knowledge. Can it be mapped? Research by Heidi Williams (MIT) does this by using evidence from the human genome project (HGP) to examine the role of intellectual property rights (IPR) on innovation. During the final years of the HGP, Celera Corporation was granted temporary licensing rights for sequences they identified prior to HGP coverage. Williams (2013) took advantage of this natural experiment to investigate whether follow-on research on individual genes in the post-HGP era was impacted by Celera's IPR claims. Her results suggest an approximately 30 percent reduction in subsequent publications, phenotype-genotype linkages, and diagnostic tests for genes first sequenced by Celera—the company's licensing rights impacted cumulative knowledge occurring at the level of a research community.

Turning to the topic of uncertainty, and the highly skewed nature of knowledge creation and innovation, Stern highlighted three points. First, he cited the work of Uzzi et al. (2013) who used unconventional metrics to predict which publications become the rare ones that go on to have a high impact. Their analysis of 17.9 million papers from all scientific fields suggests that science follows (p. 1) "a nearly universal pattern: the highest-impact science is primarily grounded in exceptionally conventional combinations of prior work yet that simultaneously features an intrusion of unusual combinations." Stern called insights from these kinds of analyses "game changers." In a similar fashion, the research by Cathy Fazio and her colleagues (summarized in Chapter 6) provided insights about the structure of skew in the area of entrepreneurial quality, where a high portion of consequential outcomes emerge from 1 or even one-tenth of 1 percent of the overall activity.

Stern turned next to the question of how the skewed impact of patents and other phenomena related to innovation can be mapped in a way that captures cumulativeness, uncertainty, and high skew. He noted the fact that discovery may occur in an area that was unanticipated. By way of analogy, Stern described how the availability of "open" satellite image maps of the Earth impacted discovery and entrepreneurship in the gold industry. Research by Nagaraj (2015) found a large and persistent difference in the rate of gold mining and discovery depending on the availability of public images—open-access maps—for a given geographical area. Furthermore, entrepreneurs were found to be far more likely to take advantage of open-access maps than established firms. Providing the mechanism for this application was a research community who figured out that these images could dramatically improve the chances of understanding where gold was located, which had nothing to do with the initial motivations for the NASA Landsat Program. The takeaway is that discovery of gold, which is unexpected and highly skewed (but concentrated in regions where the satellite imagery was available), benefited from research developed for other purposes.

Stern concluded with several statements about future measurement and policy making:

- Innovation statistics and metrics are increasingly being used to evaluate and track innovation systems at multiple levels of granularity.
- There is a need to develop meaningful connections that allow cohesive assessment of the role of different elements of the innovation system (inputs and outputs) over time. This will not always require new data sources. It may be a matter of connecting existing data—and combining traditional measures with alternatives from unanticipated sources—in creative ways.
- There is a need in measurement frameworks to recognize cumulativeness, uncertain and highly skewed phenomena, and distributed impacts.

This agenda, he argued, is particularly important for areas of innovation beyond traditional "tech to market" applications of science, as it relates to emerging uses of digital knowledge, maps, and nonscience knowledge systems.

LINKING PATENT METRICS TO OTHER INDICATORS

Jeff Oldham (Google) presented insights that can be drawn about innovation from linking patent metrics, which summarize sets of patents, to other types of indicators. Following up on one of Stern's points, Oldham noted that one way to make knowledge more cumulative is to increase the accessibility of statistics and indicators. This is analogous to the idea behind the biological resource centers that Stern discussed.

The key to increasing the value of patent data is to create the capacity to link them to a web of other innovative indicators, Oldham said. In this way, patent data can serve as a nucleus of innovation measurement. Currently, anyone can go to Google Patents or to the U.S. Patent and Trademark Office (USPTO) Website to research detailed information about a patent; there are low barriers of entry for this kind of inquiry. But it is difficult to compute over collections of patent cases. This is why patent metrics, based on summaries, have been developed.

Among the metrics that can be generated from patent summaries are publication numbers (which serve as keys), applications and grants, scope of patents by CPC codes,[1] family members (whose patents can be linked), backward and forward citations, and publication dates. A wide range of patent statistics can be added. Researchers can also compute their own patent statistics using simple structured query language (SQL) queries.

Oldham provided one sample application—an analysis of how long it takes for patents to be approved by the USPTO and whether, over a period of time, that process has slowed or sped up. Estimates from this analysis reveal that, for example, sometime between 2000 and 2001, the process grew much faster, and it has continued on this trend up until now.

After providing a quick overview of what can be done with patent metrics, Oldham turned to his main message: The analytic power of patent metrics can be greatly increased by linking them to other databases—in other words, by making patent metrics the core of a web of innovative indicators. Specifically, Oldham suggested that links to the following sources would create analytic content:

- census data—to assess patents' impact on people's quality of life;
- inventor school(s)—to assess training effectiveness;
- inventor school funding sources—to inform STAR METRICS;[2]
- government R&D funding—to assess program effectiveness;
- company sales data—to investigate the financial impact of design patents (Bascavusoglu-Moreau and Tether, 2011);
- sector sales data—to assess impact of innovation on sales; and
- inventor citizenship—to assess the effect of immigration policy.

Oldham concluded by describing Google's capacity to make various data sources available at low cost. For example, the company's BigQuery is basically an SQL that allows researchers to analyze patent metrics data either from a browser or programmatically. Patent metrics provide a means for summarizing patents that allows anyone to access and analyze them. Oldham expressed the desire that these tools be made available in a format that is linkable to other data sources, such as tables created by other researchers, government agencies, or companies.

During open discussion, Javier Miranda asked if Google had begun

---

[1]CPC stands for Cooperative Patent Classification, which is a set of codes developed jointly by the European Patent Office and the USPTO. See http://www.uspto.gov/patents-application-process/patent-search/classification-standards-and-development [August 2016].

[2]From https://www.starmetrics.nih.gov/ [August 2016]: "STAR METRICS® is a federal and research institution collaboration to create a repository of data and tools that will be useful to assess the impact of federal R&D investments. The National Institutes of Health (NIH) and the National Science Foundation (NSF), under the auspices of Office of Science and Technology Policy (OSTP), are leading this project."

to think through how data could be combined in safe ways so that information could be extracted while still residing in various silos. Oldham responded that cloud data systems have access-control levels, so it is possible to provide access to approved people for specific projects. He also noted the promise of remote access protocols whereby certified individuals from within an organization run programs and deliver the results to external researchers, or that researchers could compute over particular fields or experiment with a sample.

Sallie Keller (Social and Decision Analytics Laboratory) agreed that technological solutions will continue to emerge. She pointed to progress in such fields as the biosciences around genomics and related areas. Computer scientists in the cryptography world are working on access and confidentiality issues and finding that it is possible to compute over two distinct datasets that do not leave their silos. She characterized them as solvable problems 5 or 10 years from now. And, finally, while the concept of data governance is central across statistical agencies, she argued that it is important to begin thinking about changing the conversation about what privacy and confidentiality mean.

INNOVATION INITIATIVES AT THE STATISTICAL AGENCIES

Ron Jarmin (U.S. Census Bureau) discussed activities and plans of U.S. statistical agencies to continue improvement of their data programs in the areas of business dynamics and innovation. Part of the plan calls for increased collaboration among such agencies as the Census Bureau, Bureau of Economic Analysis, and Bureau of Labor Statistics. Jarmin's remarks focused on an initiative at the Census Bureau directed at measuring the inputs and outputs of innovation, or the activities that may lead to innovation, on a granular level. It is a collaborative research project among the Census Bureau, University of Michigan, Ohio State University, University of Chicago, and New York University that links administrative data from these institutions on funded research projects with data assets at the Census Bureau.

Jarmin said the agency's goals for the program are to: (1) improve measurement of a small but important sector of the U.S. economy—individuals involved with funded research grants; (2) address data gaps in the measurement of innovation and its relation to economic growth; (3) collaborate with data providers to deliver data products they value, such as customized reports; and (4) initiate a prototype project that can be scaled and extended to other sectors of the economy. All of this is consistent with the Census Bureau's economic and social measurement mission and directly relevant to the data providers, he said.

"Mashing up" university administrative and Census Bureau data to create new statistics and facilitate research requires collaboration, in this case, with the University of Michigan's Institute on Research in Innovation and Science (IRIS). Data from IRIS-UMETRICS on sponsored research projects and the faculty, staff, postdocs, and students involved as grant recipients (see Jason Owen-Smith's discussion in Chapter 4) allows the program to experiment with using "fat pipe" data for a sector of the economy. These data are highly complementary to business and household data at the Census Bureau, including the Business Register, Person Identification Validation System, Longitudinal Business Database, and the Longitudinal Employer-Household Dynamics (LEHD) Program underlying the Quarterly Workforce Indicators.

Linking these data, analysts at the Census Bureau are able to identify individuals and link them to records in the LEHD infrastructure, which includes data on all jobs in the economy covered by state unemployment insurance programs. Data on transactions can be linked to equipment and supply purchases and to the Business Register in a way that allows researchers to analyze the upstream and downstream value chain of university-based research. Accurate linkage is one of the major challenges inherent in the project. The quality of the links is directly proportional to the quality of the input data. Universities so far have not provided the Census Bureau with individual identifiers such as Social Security numbers that would make linking more accurate; name, address, and sometimes date of birth fields may be used.

The link to census data allows researchers to investigate some of the flows associated with science investments to universities. Hiring and spending can be tracked at a granular level. Even detailed transactions, such as where universities are sourcing their laboratory mice, can be tracked at the lab level. More generally, one can determine the locations of businesses that provide materials purchased by the universities. From employment records, the flow of human capital (mainly students, but some faculty and staff as well) from the university outward into the economy can also be tracked. These data can address the question about what happens to people involved in funded research when they leave the university, Jarmin noted. For example, taxpayers in Michigan can see that more graduates who received grants on projects took jobs in Michigan than anywhere else. The characteristics of the hiring companies can also be tracked, which is helpful for creating a picture of the health and direction of local economies.

Jarmin pointed out that this project is only scratching the surface of what can be done with linked datasets; the ability to track the inputs and the outputs at a granular level opens a wide range of possibilities. Similar programs could in principle be set up for private-sector organizations that would allow tracking of upstream and downstream impacts of a broader array of activities in the economy than university-funded research. Systematizing standards of how companies might implement such data collections to bring in economy-wide transactions would be a difficult task. But, Jarmin suggested, if there were particular areas of interest such as the health care sector where advances in electronic health records are creating new possibilities, it could be possible to expand in new directions.

Jarmin concluded with the observation that economists have always wanted access to data on every transaction in the economy. The fact that it is now possible to take a group of organizations and look at every transaction for a particular set of activities shows promise that these desires may become reality. It may never be practical to measure every transaction in the economy, he said, but there is a vast space between the present and the near future in this regard.

LEVERAGING NONSURVEY, LOCAL DATA SOURCES

Sallie Keller (Social and Decision Analytics Laboratory) pulled together a number of discussion threads developed over the course of the workshop about combining alternative types of data to analyze and understand society. She focused on leveraging local data sources in new and foundational ways. Nonsurvey data—some of it collected in real time while individuals are engaged in day-to-day life situations—provide a new lens on social observation, she said. Much of her work is oriented toward adapting statistical methods to make the best use of these data. As with high-powered telescopes peering into the universe, these emerging data sources may not be exposing things that are new, but they are exposing things that we have not been able to see before.

Eschewing the term "big data," Keller described the data revolution in more precise categorical terms. Specifically, data analysts are now able to call on designed data collections (statistically designed and intentional observational data collections); administrative data (data collected for the administration of an organization or program); opportunity data (data generated through daily activities); and procedural data (data derived from policies and procedures).

Designed data collection includes the surveys that statistical agencies have conducted for decades, which Keller said are very good in terms of quality control and application. Administrative data are increasingly important in that they are broadly collected and highly underutilized across many areas of science and evidence-based policy. Opportunity data exist on the Internet and other platforms (e.g., sensors, GPS, and video) that can be captured and analyzed. These emerging data sources do not make standard ways of observing or collecting data obsolete,

Keller argued, but they create opportunities to observe behavior at a finer level of granularity than with designed and administrative data. Finally, she said, procedural data provide information about policies and practices that govern a variety of activities and that provide the context necessary for comprehensive studies of various phenomena.

In addition to their advantages, new sources of data and new methods of combining them also create major challenges. Keller suggested that, while researchers innately address these challenges as they use data, as a community, there needs to be coordinated, disciplined ways of approaching them. The challenge with using such a diverse data landscape is data quality—both within and across buckets. The traditional approach in science involves controlling measurement processes, whether with a survey or a physical bench lab instrument. Disciplines have gone through great pains to develop theory and statistical methods—optimizing sampling frames and designing collection processes in engineering and in bench sciences. Considerations of data quality have also required controlling the ownership of data to ensure that reliability and quality are maintained. In three of the four data buckets, Keller pointed out the lack of control over data quality compared with traditional survey data. Because administrative data are collected primarily for nonresearch uses, they have to be repurposed to be used in statistical models and analyses. Another major challenge to using new kinds of data—and one she noted was not addressed extensively during the workshop—is privacy and confidentiality. Not all work can be done in secure research data centers, so this will continue to be an issue for some time.

Keller walked through some case studies being done by her lab for the Census Bureau that leverage new and detailed local data to support or enhance, or even possibly replace, some of the federal reporting and in particular some of the estimates in the American Community Survey (ACS). The idea underlying the projects, she said, is to "open the aperture" to all possible data sources, not just the ones that are easy to access.

In an education-sector case study, Keller's group set out to acquire sources of data for key variables:

1. Profiled variables: student ID, district code, year, gender, race/ethnicity, grade, age, and other variables (e.g., limited English proficiency)—data for most of these variables were complete and consistent requiring very little cleaning.
2. Transformed variables: school districts were matched with counties; ages were calculated from birthdates (North Carolina and Kentucky); for Texas, enrollment estimates were weighted to match the state level counts.

3. Restructured variables: Virginia data were restructured to create tables for race/ethnicity, by grade, gender by grade, and disadvantaged status by grade.

Figuring out which data are relevant and useful for the question at hand requires careful planning, Keller said. For state longitudinal data systems, challenges exist with geographic alignment whenever local data are being combined with state or federal sources. She noted that Feldman's Research Triangle Park account provided a good example of this challenge as the Raleigh-Durham area spans two Metropolitan Statistical Areas (see Chapter 6). Likewise, school districts do not necessarily align with the Census Bureau's Public Use Microdata Areas or even with counties. But, Keller argued, these are solvable problems.

Because the data assembled by Keller's team are so much more granular than traditional federal statistical agency sources, it becomes possible to analyze new things. For example, in an analysis of school dropout data in Kentucky, the team was able to map rates for males and females in the Appalachian and non-Appalachian regions of the state in correspondence with the reasons for dropping out—including pull factors (e.g., employment, illness, pregnancy) and push factors (e.g., failing classes, expulsion).

Similarly, their analyses of housing exploited detailed data, including local tax assessment data, data from a host of vendors, a number of transformed variables (e.g., parcels weighted by estimated number of units), and restructured variables to create consistent geocodes per parcel. These data were used to produce neighborhood profiles at highly localized levels of geography. From the ACS, one can estimate median house values across census tracts. With local data, it becomes possible to compute actual house values and, if housing value is a surrogate for wealth, a granular index on wealth diversity at the neighborhood level.

Keller offered the following concluding points. First, she said, the use of external data (especially private-sector data) requires understanding and mitigating problems that may exist with the data because analysts and policy makers have no control over their collection; this is the opposite of the federal statistics paradigm developed over the course of decades. Second, case studies are extremely helpful for developing and testing new kinds of data frameworks. Finally, a disciplined, yet flexible and adaptable, data framework is needed to assess data quality and fitness-for-use. Keller said she envisions that analytic use of granular data from multiple sources, including data quality assessments, can be done in a disciplined way such that students in 10 years will have an approach for administrative, opportunity, and procedural data that is similarly rigorous to that which is currently in place for survey data.

8

Key Themes and Possible Next Steps

Scott Stern and Bronwyn Hall closed the workshop with comments on what they saw as some of the main points presented and discussed. Stern noted three broad questions were addressed, in some cases directly and in others indirectly:

1. What can be done to facilitate productive updating of the *Oslo Manual*, which defines innovation and prescribes an approach to its measurement, going forward?
2. How should metrics be developed and modernized to capture the components of innovation—including its inputs, outputs, and dynamics over time—more accurately and more comprehensively?
3. How are new (or revised) measures and datasets changing the understanding of innovation?

Stern noted that an objective of the workshop organizing committee was to envision how the range of measures could potentially be improved and expanded to more fully capture, on a reliable basis, aspects of innovation processes that are currently missed or not completely measured. This goal requires first taking stock of the work already being done by the National Center for Science and Engineering Statistics (NCSES) and other statistical agencies. The value of their contributions, made possible by significant public investments, improve the understanding of innovation and advance its measurement and was clearly evident at many steps along the way during the proceedings. While highlighting the value for new kinds of data and new analytic constructs, Stern credited workshop participants for recognizing that the current apparatus—for example, the Community Innovation Survey (CIS) and the Business R&D and Innovation Survey (BRDIS)—contains invaluable tools that benefit from a balance between standardization (across countries and time) and experimentation (to uncover new phenomena and local circumstance). Stern made the point that planners should be explicit about what parts of a survey and resultant data products require standardization (because the capacity to make direct comparisons across countries and time has value) and what parts could be opened up to experimentation.

While progress in the field was acknowledged, several workshop participants sought to identify what Ben Martin (Chapter 2) termed the "dark matter"—aspects of innovation that are not measured and, consequently, not well understood. Over the course of the presentations, Stern observed, it became clear that shedding light on this dark matter will entail individual and organizational initiatives on domains outside the current scope of the *Oslo Manual* and the CIS upon which it is based. In some cases, the comparative advantage for advancing measurement will be in collaboration with the statistical agencies, while in other cases, researchers will need to take the lead.

Regarding revision of the *Oslo Manual*, Stern pointed out that the definition of innovation matters. Fred Gault (Chapter 2) commented that the definition is central because it is the anchor determining what is in scope and what is out of scope. Thus, to some extent, a narrow definition will confine inquiry and a broad definition will stimulate wider inquiry.

A major theme that emerged from this workshop—reiterated in the presentation by Hall—was that, to be useful, any framework must be capable of reflecting the complex and diverse processes and interactions inherent in innovation:

> Science, technology, and innovation consist of a number of components linked by the knowledge and resources that flow among them. These components include governments, government research laboratories and extension services, the intellectual property system, higher education and research institutions, venture capital, and industrial research laboratories. Perhaps less obviously, they also include individuals across many economic sectors who are engaged in improving the efficiency of production, introducing new ideas and new products, or attempting to adopt new technologies or methods of organization. When the system works well, the interaction of these institutions and individuals [turning ideas into value] produces welfare-enhancing economic growth via the introduction of new and improved processes, products, and services (Hall and Jaffe, 2012, p. 34).

Ideally, Hall argued, a measurement framework would specify a system of indicators that, taken together, are helpful in monitoring these processes and informing policies designed to foster, facilitate, and enhance them. The conceptual framework drives (implicitly or explicitly) data collection and data organization, as well as what measurements based on those data will be illuminating. Subsequently, when measurements are used in research, a feedback mechanism is created that further shapes the framework and helps to identify data and indicator gaps that, if filled, could help answer key questions about the pace, location, and nature of innovation activity (Hall and Jaffe, 2012).

Though the idea of a single unified indicator framework has allure, it is not possible if the goal is to promote research on the many ways that innovation is fostered and impacts the economy and society. This point was raised in the presentations during the workshop session on "regional innovation models and data needs." At subnational levels of analysis, it becomes especially clear that multiple approaches are needed to serve the many kinds of users of innovation (and science and technology) indicators and to address the wide range of policy questions posed. No one set or system of indicators will suit all monitoring, benchmarking, evaluating, and forecasting purposes—a unified framework is probably not a realistic goal. The range of objectives across stakeholders represented at the workshop dictates that multiple approaches may be considered when designing data collection and indicator development programs. For example, as noted above, since some potential benefits of innovation—such as better health, greater happiness, and improved environment—are not market based, other concepts of output (and sets of indicators to capture their trends) are also needed.

Accounts of innovation are also increasingly capturing a broader range of sources and administrative units. As discussed by Owen-Smith, von Hippel, and Winston Smith, among others (see Chapters 5 and 6), innovation can be tracked at the industry and sectoral levels but, in principle, it can also be traced back to where it originates, which may be businesses, households, individual entrepreneurs, universities, nonprofit organizations, or the public sector (or some combination of these). The *Oslo Manual* was created largely for the production of national-level and industry-level innovation statistics, which are important. Yet several presenters (see Chapter 6) demonstrated that measurement can be applied with illuminating effect at more localized levels—levels that are often more relevant in terms of understanding the sources of innovation (such as the human capital and geographic clustering of activities) and their impact on economic and social outcomes.

Additionally, measurement frameworks should reflect linkages in innovation processes so that flows of inputs, outcomes, and broader economic and societal impacts can be tracked. Charles Edquist (Chapter 5) noted the importance of distinguishing innovations from their determinants. He viewed entrepreneurship and research, for example, as drivers of innovation. He said these determinants need to be measured, as do the innovations themselves and the impact they have, in a detailed and coordinated way.

Regarding modernizing innovation measurement and metrics, Edquist reiterated that the scope of data collection must continue to broaden beyond R&D and patents. Stern noted the modal notion is that R&D is undertaken in a company, which then leverages the knowledge generated to produce one or more innovations—suggesting an alignment between R&D as the input with the output being new or significantly changed goods and services produced by the company. This perception may have been sufficiently accurate at the time when a number of key surveys were designed, but the interaction of different players is now more complex. To portray innovation fully, data systems should be capable of detecting knowledge linkages between people and companies—and business and household interactions generally—over time (as noted by Javier Miranda, Chapter 3), and the division of labor, which means that people engaged in idea development are often different than those commercializing the innovations (Wesley Cohen and Ashish Arora, Chapter 3). In addition, Stern concluded that the role of "nontraditional" but prominent aspects of value creation, such as design (Bruce Tether, Chapter 3), digitization, and robotics (Rob Seamans, Chapter 3) is necessary. These complexities are only compounded by the changing innovation landscape that includes ever-increasing internationalization, globalization, networking, and co-invention. Similarly, Robbins (2016, p. 11) in the background paper prepared for the workshop noted, "the increasingly open nature of innovation, including the role of suppliers, customers, licensed technology, and collaborations and joint projects, suggest that the activities of firms need to be understood in the context of their relationships to other institutions, including other firms, universities, non-profits, and government entities."

In her summary comments, Hall noted that a comprehensive portrayal requires acknowledging that innovators (and not just the innovations they produce) matter: that is, a recognition that knowledge flows via individuals. Innovation is undertaken by a person or teams of persons, which makes maintaining individual identifiers in datasets crucial to tracking these linkages and for anticipating where action may occur. This point was made by Eric von Hippel (Chapter 6), Sallie Keller (Chapter 7), and others: People are innovators. Scientists, engineers, and designers, as well as people in the broader community, can be identified as engaging in a distinctive human trait, which is making tools, using tools, problem solving, and developing novel creations. Winston Smith commented that this notion of tracking innovators as well as innovations is important at both the level of measurement (e.g., suggesting a need for data on the migration of science, technology, engineering, and mathematics graduates to firms) and at the level of the questions asked.

Ben Martin (Chapter 2) reinforced discussion to move beyond traditional indicators. In particular, he argued that indicators based on patent data are often only distant proxies for the true measurement objective. Jeff Oldham (Chapter 7) also demonstrated that patent data sources could be used more effectively, by linking them to nonpatent data, to shed light on the impact of patented ideas on sales, quality of life, educational effectiveness, and other areas. Several participants argued it was time to broaden sectoral coverage of innovation. Services and experiences increasingly dominate value-added and employment in the United States and other advanced economies, they pointed out, but there is still a manufacturing orientation in many measures of innovation. Martin noted, for example, that if more had been known about the characteristics of innovation in the financial sector, policy makers and researchers would have been better placed to anticipate the events leading to the 2007 crisis. This also highlights that not all innovation is socially desirable.

Several presentations highlighted that comprehensive conceptualization and measurement of innovation also requires recognizing activities (and outcomes) that take place in households (Eric von Hippel, Chapter 5), in the public and nonprofit sectors (Kay Husbands Fealing, Chapter 5), and in communities (all presenters in Chapter 6). Stern remarked that a goal for the future should be to better capture—in both survey-based and administrative datasets as well as the broader programs of the National Center for Science and Engineering Statistics, OECD, and others—the scope of innovation in a way that integrates leading-edge research findings (which breaks new ground) into the statistical machinery (which facilitates consensus and common understanding).

Including the public sector in "official" innovation measurement raises an important issue (which also applies to some other sectors), which is the dearth of data on outcomes. Hall pointed out that measurement of outcomes for public-sector innovation requires thinking about how survey questions should be asked. The wording of survey instruments has to change not only for the public sector and not only for the innovation question, but also possibly for other questions, she noted. Measurement of the innovation and public goods link (which may take place in the public or private sector) requires connecting indicators of the inputs (mainly dollars and human capital) all the way through to social outcomes in the ecosystems in which they take place. As von Hippel noted (Chapter 5), much of the current focus is on economic outcomes or tangible outputs, which leads to an overemphasis on patents and other readily available data; many of the useful innovations that bring high utility for the public good are not "patented ideas."

The development of innovation measurement will, to a great degree, be driven by the availability of new kinds of data—where, as Stern phrased it, the administrative data revolution meets the digitization revolution. The administrative data revolution discussed by Javier Miranda (Chapter 3) reflects at least two developments. One is an increasing capacity at the statistical agencies and others to link different administrative and survey data sources; Stern cited the work by Chetty et al. (2014) combining IRS earnings records and other sources to examine determinants of intergenerational economic mobility as one high-profile example. Elsewhere, there are data on entrepreneurship, around health records, around studies of schools and teacher value added, and others. In addition, Stern observed, the digitization revolution allows researchers to uncover new ways to understand company activities, new products, and new services that reflect the process of innovation. Except for a few examples, these two developments have so far existed as separate tracks within the innovation community. Sallie Keller (Chapter 7) provided an overview of the role of nonsurvey data for statistical use, and of how statistical and data-gathering techniques are already being used to advance research interests.

As articulated by Robbins (2016), innovation-related data priorities for the statistical agencies can be informed by evaluating the needs of the data-using community against three factors: (1) data currently available from NCSES and other statistical agencies, (2) increasing opportunities provided by nonsurvey data, and (3) opportunities for linking existing datasets. Further, capabilities of the data-collecting agencies can be enhanced for these priorities by leveraging partnerships, collaborations, and supporting research. Several participants identified enhancing the ability to link data across sources as a prerequisite to analytic and measurement progress. One example among the many kinds of new data that will aid a greater understanding of innovation are those tracking angel funding and crowd funding, but currently, the data are still not sufficiently detailed and available to enable sustained high-quality research on new innovation funding.

Regarding what has been learned and continuing gaps, Stern observed that researchers are able to measure some aspects of innovation extremely well. However, he said, the share of what is being missed—the portion identified by Martin's dark matter analogy—is still largely unknown, except perhaps in a highly aggregated sense as reflected in statistics on productivity (as described by Dan Sichel, Chapter 3). And productivity measurement is essentially a black box—a residual measure of what is not explained by changes in the levels of inputs of production. Even assessing the amount missed on the basis of productivity measures (without claiming to identify the sources) is controversial, Stern noted, as it requires anchoring one's belief that the input-output relationships are captured accurately, which is not universally accepted.

Stern described the following areas for future research identified by workshop participants as key to shedding light on some of the innovation "dark matter":

- The division of innovative labor: Research by Cohen, Arora, and others is demonstrating that a surprisingly high fraction of important innovations come from outside the firm, albeit sometimes cooperatively, Stern said. The linkages among innovating individuals and organizations need to be better understood.
- Design, digitization, robotics, household users and entrepreneurship: Workshop discussants (specifically, Tether, Chapter 3) noted that aspects of innovation such as design could be systematized enough to be measured at the level of a "design index." Rob Seamans' presentation (Chapter 3) on innovations in the area of automation illustrated that there is not a single use dataset that currently captures job-creating and job-replacing phenomena adequately.
- Centrality of the regional systems approach: A considerable amount of new work is seeking to understand entrepreneurship and other innovative activities at the regional level. Practitioners and policy makers have become accustomed to the idea that innovation statistics are produced at the national level. In the 1990s, outside of a few state-level aggregates, science and engineering indicators were based at the level of countries. Stern observed that at least in a large country such as the United States, many processes can be better understood at the regional level. Presentations by Feldman, Fazio, Ziedonis, and Winston Smith (Chapter 6) made it clear that there is probably more to be learned at the subnational unit of analysis than at the national level, especially because both research communities and policy communities are often regional in character.

The last area that Stern noted as emerging from the workshop concerned the skewed distributions of discovery, invention, innovation, and entrepreneurship and the implications for data and research. Stern referred to an empirical phenomenon that a large percentage of profoundly influential innovative activity—for example, returns on initial public offerings or the impact of ideas on profits, products, and job creation—occurs from a small proportion of the distribution. This means that, for some measurement purposes, detailed statistics based on 1 percent of activity (if it is the right 1%, or perhaps 2 or 3%) are very useful, while information on the full distribution of total number of papers or patents or research and development that a country or a region produced is inefficient in producing insights. At an aggregate level, he said, it may be more important to know if a country or a region is reliably producing innovations that appear in the top 0.5 percent of the distribution measured along dimension x than to know their totals. An important research priority for the near-term horizon, then, is to evaluate what statistics can be produced at the nth percentile as part of the information infrastructure. The Census Bureau's Business Dynamics Statistics group has done pioneering work to reveal the analogous skewedness of job creation and destruction, firm survival, and other areas. Stern suggested that this work might provide a blueprint for further work on the topic. Just as focusing on the top 1 percent in the area of income changed the conversation of what people studied, focusing more centrally on these skewed distributions in areas related to innovation is important for advancing understanding of the relationship between innovation and its sources and outcomes.

References

Aghion, P., and Howitt, P. (1998). *Endogenous Growth Theory*. Cambridge, MA: MIT Press.

Amabile, T. (1996). *Creativity in Context*. Boulder, CO: Westview Press.

Angel Resource Institute. (2014). Angel group update: 2014, *The Halo Report™*. See https://www.angelcapitalassociation.org/data/Documents/Resources/ARIHaloReport2014FINAL.pdf [August 2016].

Arora, A., Cohen, W.M., and Walsh, J.P. (2016a). The acquisition and commercialization of invention in American manufacturing: Incidence and impact. *Research Policy*, 45(6), 1113-1128.

Arora, A., Cohen, W.M., and Walsh, J.P. (2016b). Innovation and the sources of invention in American manufacturing. Submitted to *Research Policy*.

Asheim, B., and Coenen, L. (2005). Knowledge bases and regional innovation systems: Comparing Nordic clusters. *Research Policy, 34*, 1173-1190. doi:10.1016/j.respol.2005.03.013.

Azoulay, P., Liu, C.C., and Stuart, T.E. (2011). *Social Influence Given (partially) Deliberate Matching: Career Imprints in the Creation of Academic Entrepreneurs*. Working Paper. Cambridge, MA: Harvard University.

Balsmeier, B., Fleming, L., and Manso, G. (Forthcoming). Independent boards and innovation. *Journal of Financial Economics*.

Bartling, B., Ernst, F., and Holger, H. (2014, June). *The Intrinsic Value of Decision Rights*. University of Zurich, Department of Economics Working Paper No. 120. Available: https://ssrn.com/abstract=2255992 [August 2016].

Bascavusoglu-Moreau, E., and Tether, B. (2011). *Design Economics Chapter Two: Registered Designs & Business Performance—Exploring the Link*. Report commissioned for the Intellectual Property Office, United Kingdom. Available: https://www.gov.uk/government/uploads/system/uploads/attachment_data/file/310638/ipresearch-designsreport2-201109.pdf [August 2016].

Bertrand, M., and Schoar, A. (2003). Managing with style: The effect of managers on firm policies. *Quarterly Journal of Economics, 118*(4), 1169-1208.

Bloch, C., and Bugge, M.M. (2013). Public sector innovation—From theory to measurement. *Structural Change and Economic Dynamics, Elsevier, 27*(C), 133-145.

Bloom, N., and Van Reenen, J. (2007). Measuring and explaining management practices across firms and countries. *Quarterly Journal of Economics, 122*(4), 1351-1408.

Bloom, N., Sadun, R., and Van Reenan, J. (2012, February). Americans do it better: U.S. multinationals and the productivity miracle. *American Economic Review, 102*(1), 167-201. doi: 10.1257/aer.102.1.167.

Byrne, D.M., Oliner, S.D., and Sichel, D.E. (2013). Is the information technology revolution over? *International Productivity Monitor, 25*, 20-36.

Byrne, D.M., Oliner, S.D., and Sichel, D.E. (2015). *How Fast Are Semiconductor Prices Falling?* AEI Working Paper. Washington, DC: American Enterprise Institute.

Chetty, R., Hendren, N., Kline, P., Saez, E, and Turner, N. (2014, January). *Is the United States Still a Land Of Opportunity? Recent Trends in Intergenerational Mobility*. NBER Working Paper No. 19844. Cambridge, MA: National Bureau of Economic Research.

Corrado, C., Goodridge, P., and Haskel, J. (2011). *Constructing a Price Deflator for R&D: Calculating the Price of Knowledge Investments as a Residual*. Discussion Paper, No. 2011/07. London, UK: Imperial College.

Cutler, D.M., and McClellan, M. (2001). Is technological change in medicine worth it? *Health Affairs, 20*, 11-29.

Dasgupta, P., and David, P. (1994). Towards a new economics of science. *Research Policy, 23*(5), 487-521.

DeBruyn, J. (2015). GSK's Triangle genealogy. *Triangle Business Journal, 3*(1), September 4.

Decker, R., Haltiwanger, J., Jarmin, R., and Miranda, J. (2015, March). *Where Has All the Skewness Gone? The Decline in High Growth (Young) Firms in the U.S.* Available: http://econweb.umd.edu/~decker/DHJM_3_30_2015.pdf [August 2016].

Ederer, F., and Manso, G. (2013). Is pay for performance detrimental to innovation? *Management Science, 59*, 1496-1513.

Edquist, C., and Zabala-Iturriagagoitia, J.M. (2015). *The Innovation Union Scoreboard Is Flawed: The Case of Sweden—Not the Innovation Leader of the EU—Updated Version*. CIRCLE Working Paper 2015/27. Lund, Sweden: Center for Innovation, Research, and Competence in the Learning Economy, Lund University.

Fazio, C., Guzman, J., Murray, F., and Stern, S. (2016). *A New View of the Skew: A Quantitative Assessment of the Quality of American Entrepreneurship*. Cambridge, MA: MIT Innovation Initiative, Laboratory for Innovation Science and Policy. Available: http://innovation.mit.edu/assets/A-New-View_Final-Report_5.4.16.pdf [August 2016].

Feldman, M., Hadjimichael, T., and Lanahan, L. (2016). The logic of economic development: A definition and model for investment. *Environment and Planning C: Government and Policy, 34*(1), 5-21. doi: 10.1177/0263774X15614653.

Fernald, J.G. (2012). *A Quarterly, Utilization-Adjusted Series on Total Factor Productivity*. FRBSF Working Paper, No. 2012-19. San Francisco, CA: Federal Reserve Board.

Fernald, J., and Jones, C. (2014). *The Future of U.S. Economic Growth. AEA Papers and Proceedings*. Washington, DC: American Economic Association.

Frey, C.B., and Osbourne, M.A. (2013, September). *The Future of Employment: How Susceptible Are Jobs to Computerisation?* Working Paper, Oxford Martin Programme on Technology and Employment, University of Oxford. Available: http://www.oxfordmartin.ox.ac.uk/downloads/academic/future-of-employment.pdf [August 2016].

Furman, J.L., and Stern, S. (2011). Climbing atop the shoulders of giants: The impact of institutions on cumulative research. *American Economic Review, 101*, 1933-1963. doi:10.1257/aer.101.5.1933.

Gagné, M., and Deci, E.L. (2005). Self-determination theory and work motivation. *Journal of Organizational Behavior, 26*, 331-362. doi:10.1002/job.322.

Galindo-Rueda, F., and Millot, V. (2015). *Measuring Design and Its Role in Innovation*. OECD Science, Technology, and Industry Working Paper No. 2015/01. Paris, France: OECD.

Gambardella, A., Khashabi, P., and Panico, C., (2013, November). *Working Autonomy in Innovative Activities: Managing Knowledge Workers*. Academy of Management Annual Meeting Proceedings. Washington, DC. Available: http://proceedings.aom.org/content/2013/1/14788.short [November 2016].

Gambardella, A., Raasch, C., and von Hippel, E. (2015). *The User Innovation Paradigm: Impacts on Markets and Welfare*. Social Science Research Network. Available: http://papers.ssrn.com/sol3/papers.cfm?abstract_id=2079763 [August 2016].

Gault, F. (2016). *Measuring Innovation in All Sectors of the Economy*. United Nations University, Working Paper Series, No. 2015-038. Maastricht, The Netherlands. Available: http://www.merit.unu.edu/publications/working-papers/abstract/?id=5832 [August 2016].

Gordon, R.J. (2016). *The Rise and Fall of American Growth*. Princeton, NJ: Princeton University Press.

Graetz, G., and Michaels, G. (2015). Estimating the impact of robots on productivity and employment. *VOX*, CEPR's Policy Portal, Washington, DC. Available: http://voxeu.org/article/robots-productivity-and-jobs [August 2016].

Graham, S., Grim, C., Island, T., Marco, A., and Miranda, J. (2015). *Business Dynamics of Innovating Firms: Linking U.S. Patents with Administrative Data on Workers and Firms*. Center for Economic Studies Research Paper CES 15-19. Washington, DC: U.S. Census Bureau. Available: https://www2.census.gov/ces/wp/2015/CES-WP-15-19.pdf [August 2016].

Guzman, J., and Stern, S. (2015). Where is Silicon Valley? *Science, 347*(6222), 606-609. doi: 10.1126/science.aaa0201.

Guzman, J., and Stern, S. (2016). *The State of American Entrepreneurship: New Estimates of the Quantity and Quality of Entrepreneurship for 15 U.S. States, 1988-2014*. Working Paper No. 22095. Cambridge, MA: National Bureau of Economic Research. Available: http://www.nber.org/papers/w22095.pdf [August 2016].

Hall, B.H., and Jaffe, A.B. (2012). *Measuring Science, Technology, and Innovation: A Review*. Commissioned paper prepared for the Panel on Developing Science, Technology, and Innovation Indicators for the Future. Committee on National Statistics, Division of Behavioral and Social Sciences and Education, Washington, DC.

Haltiwanger, J., Hathaway, I., and Miranda, J. (2014). *Declining Business Dynamism in the U.S. High-Technology Sector*. Kansas City, MO: Ewing Marion Kauffman Foundation.

Hathaway, I., and Litan, R. (2014). *The Other Aging of America: The Increasing Dominance of Older Firms*. Economic Studies at Brookings. Washington, DC: The Brookings Institution. Available: https://www.brookings.edu/wp-content/uploads/2016/06/other_aging_america_dominance_older_firms_hathaway_litan.pdf [August 2016].

Hecker, D. (2005). High-technology employment: a NAICS-based update. *Monthly Labor Review (U.S. Dept. of Labor and U.S. Bureau of Labor Statistics), 128*(7), 58.

Hienerth, C., von Hippel, E., and Jensen, M. (2012). *User Community vs. Producer Innovation Development Efficiency: A First Empirical Study*. MIT Sloan Working Paper No. 4926-11, SSRN eLibrary, Social Science Research Network. Available: http://papers.ssrn.com/sol3/papers.cfm?abstract_id=1916319 [August 2016].

Holland, J.H. (2006). Studying complex adaptive systems. *Journal of Systems Science and Complexity, 19*, 1-8. Available: http://hdl.handle.net/2027.42/41486 [August 2016].

Kuhn, T. (1962). *The Structure of Scientific Revolutions*. Chicago, IL: University of Chicago Press.

Lacetera, N. (2009). Academic entrepreneurship. *Managerial and Decision Economics, 30*(7), 443-464. doi: 10.1002/mde.1461.

Lamoreaux, N., and Sokoloff, K. (1999). Inventors, firms, and the market for technology in the late nineteenth and early twentieth centuries. In N. Lamoreaux, D. Raff, and P. Temin (Eds.), *Learning by Doing in Markets, Firms, and Countries* (pp. 19-57). Chicago, IL: University of Chicago Press.

Lane, J., Owen-Smith, J., Rosen, R., and Weinberg, B. (2014). *New Linked Data on Research Investments: Scientific Workforce, Productivity, and Public Value*. NBER Working Paper No. 20683. Cambridge, MA: National Bureau of Economic Research. Available: http://www.nber.org/papers/w20683 [August 2016].

Ledford, H. (2016). CRISPR: Gene editing is just the beginning. *Nature, 531*, 156-159. doi:10.1038/531156a.

Lerner, J., and Kovner, A. (2015). Doing well by doing good? Community development venture capital. *Journal of Economics & Management Strategy, 24*(3), 643-663.

Litan, R.E. (2010). *Inventive Billion Dollar Firms: A Faster Way to Grow*. SSRN Working Paper No. 1721608. Available: https://papers.ssrn.com/sol3/papers.cfm?abstract_id=1721608 [November 2016].

Loeb, S. (2014). How successful are the top accelerators? *Financial Trends and News, Vatornews*, April 26. Available: http://vator.tv/n/363c [November 2016].

Lowe, N. (2009). Challenging tradition: Unlocking new paths to regional industrial upgrading. *Environment and Planning A, 41*(1), 128-145. doi:10.1068/a40111.

Lundvall, B. (1992). *National Innovation Systems: Towards a Theory of Innovation and Interactive Learning*. London, England: Pinter.

Manso, G. (2011). Motivating innovation. *Journal of Finance, 66*, 1823-1860.

Mason, C., and Brown, R. (2014). *Entrepreneurial Ecosystems and Growth Oriented Entrepreneurship*. Background paper prepared for the workshop organized by the OECD LEED Programme and the Dutch Ministry of Economic Affairs on Entrepreneurial Ecosystems and Growth Oriented Entrepreneurship, The Hague, Netherlands, November 2013. Final Version, January 2014. Available: https://www.oecd.org/cfe/leed/Entrepreneurial-ecosystems.pdf [August 2016].

Mokyr, J. (2004). *The Gifts of Athena: Historical Origins of the Knowledge Economy*. Princeton, NJ: Princeton University Press.

Montresor, S., and Vezzani, A. (2015). The production function of top R&D investors: Accounting for size and sector heterogeneity with quantile estimations. *Research Policy, 44*(2), 381-393.

Nagaraj, A. (2015). *The Private Impact of Public Maps—Landsat Satellite Imagery and Gold Exploration*. Job Market Paper, MIT Sloan School of Management, Cambridge, MA. Available: http://web.mit.edu/nagaraj/files/nagaraj_jmp_nov6.pdf.

National Research Council. (2014). *Capturing Change in Science, Technology, and Innovation: Improving Indicators to Inform Policy*. Panel on Developing Science, Technology, and Innovation Indicators for the Future, R.E. Litan, A.W. Wyckoff, and K.H. Husbands Fealing, Editors. Committee on National Statistics, Division of Behavioral and Social Sciences and Education. Board on Science, Technology, and Economic Policy, Division of Policy and Global Affairs. Washington, DC: The National Academies Press.

Nelson, R. (1977). *The Moon and the Ghetto*. New York: W.W. Norton.

OECD. (2007). *Science, Technology, and Innovation Indicators in a Changing World: Responding to Policy Needs*. Paris, France. Available: http://www.oecd.org/sti/inno/sciencetechnologyandinnovationindicatorsinachangingworldrespondingtopolicyneeds.htm [August 2016].

OECD. (2010). *Towards a Measurement Agenda for Innovation*. Paris, France. Available: http://www.oecd.org/site/innovationstrategy/45392693.pdf [August 2016].

OECD-Eurostat. (2005). *Oslo Manual—Guidelines for Collecting and Interpreting Innovation Data, 3rd Edition*. Paris, France. Available: http://www.oecd.org/sti/inno/oslomanualguidelinesforcollectingandinterpretinginnovationdata3rdedition.htm [August 2016].

Oppenheimer, J. (1948). The eternal optimist. *Time Magazine, 52*, 81.

Porter, M.E. (1990). *The Competitive Advantage of Nations*. Worchester, UK: Billing.

Renski, H.C., Koo, J., and Feser, E.J. (2007). Differences in labor versus value-chain industry clusters: An empirical investigation. *Growth and Change, 38*(3), 364-395.

Robb, A., and Farhat, J. (2013). *An Overview of the Kauffman Firm Survey—Results from 2011 Business Activities*. Kansas City, MO: Ewing Marion Kauffman Foundation. Available: http://www.kauffman.org/~/media/kauffman_org/research%20reports%20and%20covers/2013/06/kauffmanfirmsurvey2013.pdf [August 2016].

Robbins, C. (2016). *Concepts and Models for Focusing on People and Places in the Development of Innovation Data*. Unpublished manuscript, National Science Foundation.

Romer, P.M. (1994). The origins of endogenous growth. *The Journal of Economic Perspectives*, 8(1), 3-22. doi:10.1257/jep.8.1.3. JSTOR 2138148.

Rosenberg, N. (1982). *Inside the Black Box—Technology and Economics*. New York: Cambridge University Press.

Rosoff, M., D'Onfro, J., and Carson, B. (2015). Silicon Valley's denial is over: Everybody thinks we're in a bubble. *Business Insider*, October 10. Available: http://www.businessinsider.com/silicon-valleys-denial-is-over-everybody-thinks-were-in-a-bubble-2015-10 [August 2016].

Sander, A., and Wolfgang, M. (2014). *The Rise of Robotics*. BCG Perspectives, Center for Sensing & Mining the Future, Boston Consulting Group, Boston, MA. Available: https://www.bcgperspectives.com/content/articles/business_unit_strategy_innovation_rise_of_robotics/ [August 2016].

Sauermann, H., and Cohen, W. (2010). What makes them tick? Employee motives and firm innovation. *Management Science, 56*(12), 2134-2153.

Sauermann, H., and Roach, M. (2014). Not all scientists pay to be scientists: PhDs' preferences for publishing in industrial employment. *Research Policy, 43*(1), 32-47.

Sauermann, H., and Stephan, P. (2013). Conflicting logics? A multidimensional view of industrial and academic science. *Organization Science, 24*(3), 889-909.

Schumpeter, J.A. (1934). *The Theory of Economic Development: An Inquiry into Profits, Capital, Credit, Interest, and the Business Cycle*. Harvard University Press. Harvard Economic Studies 46. Available: http://www.hup.harvard.edu/catalog.php?isbn=9780674879904 [November 2016].

Stern, S. (2004). Do scientists pay to be scientists? *Management Science, 50*(6), 835-853.

Torrisi, S., Gambardella, A., Giuri, P., Harhoff, D., Hoisl, K., and Mariani, M. (2016). Using, blocking, and sleeping patents: Empirical evidence from a large inventor survey. *Research Policy, 45*(7), 1374-1385. Available: http://dx.doi.org/10.1016/j.respol.2016.03.021 [August 2016].

Uzzi, B., Mukherjee, S., Stringer, M., and Jones, B. (2013). Atypical combinations and scientific impact. *Science, 342*(6157), 468-472.

Van Noorden, R. (2012). Global mobility: Science on the move. *Nature, 490*(7420), News Feature, October 17. Available: http://www.nature.com/news/global-mobility-science-on-the-move-1.11602 [November 2016].

Von Hippel, E., de Jong, J., and Flowers, S. (2010, September 27). *Comparing Business and Household Sector Innovation in Consumer Products: Findings from a Representative Study in the UK*. SSRN eLibrary, Social Science Research Network. Available: http://papers.ssrn.com/sol3/papers.cfm?abstract_id=1683503 [August 2016].

Williams, H. (2013). Intellectual property rights and innovation: Evidence from the human genome. *Journal of Political Economy, 121*(1), 1-27.

Winston Smith, S., Hannigan, T.J., Gasiorowski, L. (2015). *Peering Inside: How Do Peer Effects Impact Entrepreneurial Outcomes in Accelerators?* Academy of Management Annual Meeting Proceedings. Washington, DC. Available: http://proceedings.aom.org/content/2015/1/17072 [November 2016].

Zhao, B., and Ziedonis, R. (2012, July). *State Governments as Financiers of Technology Startups: Implications for Firm Performance*. Social Science Research Network. Available: http://papers.ssrn.com/sol3/papers.cfm?abstract_id=2060739 [August 2016].

Zolas, N., Goldschlag, N., Jarmin, R., Stephan, P., Owen- Smith, J., Rosen, R.F., McFadden Allen, B., Weinberg, B.A., and Lane, J. (2015). Examining employment and earnings outcomes for Ph.D. recipients. *Science, 350*(6266), 1367-1371. Available: http://www.sciencemag.org/content/350/6266/1367.full [August 2016].

Appendix A

Workshop Agenda

**NCSES/CNSTAT Workshop on
Advancing Concepts and Models of Innovative
Activity and STI Indicator Systems**

National Academy of Sciences
2101 Constitution Ave. NW, Washington, DC 20418
Conference Room 120

Thursday-Friday, May 19-20, 2016

Workshop Goals

The workshop will bring together academic researchers, private- and public-sector experts, and public policy agencies to develop strategies for broadening and modernizing science, technology, and innovation (STI) information systems. The workshop will be oriented toward helping the NSF's National Center for Science and Engineering Statistics (NCSES) refine and prioritize its work on innovation metrics to maximize the relevance and utility of its data collection programs and statistical products to users. The focus will be on conceptualizing innovation—its inputs, outputs and consequences—in a way that reveals elements that are being measured well and those that are being measured inaccurately or not at all. Presentations and discussion will take into account the role of innovation, not only as it affects economic growth and productivity directly, but also as a mechanism for creating greater public good and meeting social challenges such as those associated economic mobility, health, civic engagement, population aging, or climate change. Workshop participants should also imagine how new kinds of information—e.g., naturally occurring, unstructured, digital data—may be used to complement more traditional survey and administrative data in the construction of innovation metrics. Participants should seek to identify questions that cannot be answered now but could be with additional data that have a reasonable chance of being collected.

Thursday, May 19 (9 am-5 pm)

9:00 am Welcome; Project Objectives

(breakfast will be available outside conference room 120)

Key issues pertaining to the changing face of innovation and potential approaches to its measurement will be identified. NCSES will provide an overview of the agency's goals and future directions, recent work, and the variety of studies initiated on innovation and human capital data. The focus on concepts and models in this workshop is meant to generate input to data development strategies that take into account both user needs and emerging opportunities.

- **Scott Stern** (MIT; ***workshop chair***): Key questions, goals for the day
- **John Gawalt** (NCSES): The scope of innovation measurement in the context of the agency's user community—e.g., for international comparisons and benchmarking, policy, academic research, business uses. Reflections on the NAS *Capturing Change* report
- **Brian Harris-Kojetin** (Deputy Director, CNSTAT): Welcome; overview of the Academies

9:15 Assessing Innovation Measurement: How Accurately Do We Measure Innovation Processes and the Resultant Outcomes Delivered to Society and the Economy?

Traditional measures of innovation, based largely on data reflecting expenditures on inputs (e.g., investment in R&D), capture the impact of innovation on economic outputs and outcomes in only a partial way. This has led for calls to collect new kinds of data and create new metrics. In this session, participants will identify aspects of the process and its impacts that are currently being measured well as well as those that are being measured incorrectly, incompletely, or not at all. Such assessment encompasses inputs to innovation (R&D and other), outputs and outcomes of innovation (economic growth, better functioning society), and the utility of metrics to stakeholders ranging from government to businesses. The takeaway from this discussion should be guidance to NSF about the extent to which the right data are being collected, and what additional kinds of information should be pursued. **[15 minutes each]**

- **Bronwyn Hall** (UC Berkeley; ***session lead***): The role and purpose of indicator frameworks in the analysis of science, technology, and innovation and economic well-being; potential impacts and relevance of extending frameworks to include public-sector innovation and broader social impacts
- **Ben Martin:** (University of Sussex): How emergence of the knowledge society shapes innovation; implications for measurement of innovation
- **Fred Gault** (UNU-MERIT, Maastricht): Developing new indicators to capture the changing nature of innovation; policy and other uses of innovation data and indicators; international comparisons
- **Fernando Galindo-Rueda** (OECD): International standards for measuring and comparing innovation; public procurement of innovation; status and overview of *Oslo Manual* revision
- **Dan Sichel** (Wellesley College): Innovation and productivity—recent puzzles

Open Discussion [15 minutes]

10:45 ***Break*** *(refreshments available outside conference room 120)*

11:00 **Innovation Beyond R&D**

Data on narrow aspects of innovation—such as R&D expenditures or patent applications, assignments, and citations—are often used to proxy levels of innovation activity. However, in the measurement community, it has long been understood that activities both technical and nontechnical, including those that are not a directly measurable function of R&D and science, also play an important role in innovation.

NCSES's Census Business R&D and Innovation Survey (BRDIS) asks firms for information about newly introduced or improved products and processes, and about purchased R&D inflows and outflows, revenue from the sale of patents and patent licensing, and a number of IP-related activities. Some questions about innovative activities can be adequately addressed with current data sources such as BRDIS, while others may require new or different kinds of data. For example, less is known about how innovation takes place in organizations that are not R&D intensive, as is often the case in the service sectors. Questions to be addressed here include: What models exist for assessing different kinds of (e.g., non-

R&D-based) innovation? What do we know about innovators acquiring inventions from external sources? How would understanding the division of innovative labor affect our understanding of the propensity to innovate? **[15 minutes each]**

- **Wesley Cohen** (Duke University; ***session lead***): Recent developments in survey-based approaches to measuring innovation, including the Community Innovation Survey
- **Ashish Arora** (Duke University): Technology licensing as an alternative to in-house R&D
- **Javier Miranda** (U.S. Census Bureau): Research using LBD/BDS and other statistical agency data that helps measure innovation.
- **Kristin McCue** (U.S. Census Bureau): New and emerging data products at statistical agencies related to innovation [Ask for documentation of current data products along with a summary of things the agencies are planning to roll out (or considering) for the future]
- **Bruce Tether** (University of Manchester): Design and development—design as an alternative or complement to R&D
- **Rob Seamans** (Council of Economic Advisers): Innovation, technology, and industrial organization; tracking automation across sectors of the economy to inform employment and other policies

Open Discussion [15 minutes]

12:30 pm ***Lunch*** *(available outside conference room 120)*

1:30 **The Role of Individuals (and networks of individuals) in Innovation**

Compared with institutions and organizations, less attention has been given to the role of individuals and teams in the innovative process. How people's education, entrepreneurial talents, and other human capital characteristics result in innovation are vast and largely unresolved research areas. One example of how innovation relates to individuals, with direct and high-profile policy content, is in the area of immigration and human capital formation. Linkages between individuals and institutions also play an important role in innovation. To what extent do we understand how these linkages to work? What opportunities to understand the role of human capital in innovation within a firm could be gained from human resource surveys and by linking existing data to other data sets? **[15 minutes each]**

- **Paula Stephan** (Georgia State University; ***session lead***): The global S&E workforce—the "Global Science" Research Project; placement of recent graduates in industry. Data needs, policy issues
- **Rajshree Agarwal** (University of Maryland): Linking knowledge diffusion among firms, industries, and regions to the underlying mechanisms of employee entrepreneurship and mobility
- **Lee Fleming** (UC Berkeley): Patent Data: New Metrics and New Linkages
- **Jason Owen-Smith** (University of Michigan): Networks, physical space, and innovation. Work on IRIS/UMETRICS to collect, improve, protect, and use big data on the dynamics of science and the economy
- **Alfonso Gambardella** (University Bocconi): Managerial practices and incentives for research and innovation; what could be learned from experiments, large and systematic computerized data bases, surveys

Open Discussion [15 minutes]

3:00 ***Break*** *(refreshments available outside conference Room 120)*

3:15 Measuring Public-Sector Innovation and Social Progress

Measuring the full range of benefits generated through innovation requires capturing data on a range of activities that goes beyond those traditionally tracked. Innovation that takes place in the public sector is one area to be explored and is a high-priority topic within the international community. Innovation that advances the public good in ways that take place beyond the market (in noneconomic ways) is another related topic of interest. Both have featured in discussions about revision of the *Oslo Manual* and in planning for September's Blue Sky Conference.

Questions to be addressed include the following: What approaches to innovation measurement can be developed for capturing activities within the public sector? Can innovation metrics be developed that bring into scope how the successful exploitation of ideas affects the provision of public goods and the well-being of society more broadly, beyond the contribution to efficiency, effectiveness, or quality in the production of market goods and services? Are there negative social impacts of innovation that should be included in our understanding of the outcomes of innovation? **[15 minutes each]**

- **Fred Gault** (UNU-MERIT, Maastricht; ***session lead***): Innovation indicators for the public sector. Conceptual overview of what would variables factor into public-sector innovation measurement
- **Daniel Sarewitz** (Arizona State University): Innovation and social progress
- **Charles Edquist** (CIRCLE, Lund University): Design of innovation policy, innovation-related public procurement and a critical assessment of the Innovation Union Scoreboard
- **Kaye Husbands Fealing** (Georgia Tech): Innovation and noneconomic/public health outcomes; GA Tech food security and safety study
- **Eric von Hippel** (MIT): Household-sector innovation

Open Discussion [15 minutes]. Overarching questions for presenters during this session and for discussion afterward:

- Which aspects of the innovation processes are not well covered and increasing in importance?
- If we could measure something that is currently not measured, what would it be? Where can funding for measurement be most productively directed?
- For what aspects of innovation measurement might surveys not be the optimal data collection strategy? Where might administrative or big data become more prominent in meeting demand for new metrics?
- What is the path forward for measuring innovation (a difficult measure process) in the case of public goods and social progress (a difficult to measure, largely nonmarket outcome)?

5:00 pm *Planned Adjournment*

Friday, May 20 (9 am-2 pm)

9:00 am **Welcome Back** *(breakfast will be available outside conference room 120)*

Scott Stern (MIT; ***workshop chair***): Key questions, goals for the day

9:15 **Regional Innovation Models and Data Needs**

Many of the spillovers and complementarities of innovative activity take place at the local level. What models help policy makers and other data users at the sub-national level? **[15 minutes each]**

- **Maryann Feldman** (UNC/NSF; ***session lead***): Regional innovation systems
- **Catherine Fazio** (MIT): Measures of innovation quantity and quality—e.g., Entrepreneurial Quality Index, the Regional Entrepreneurship Cohort Potential Index and the Regional Entrepreneurship Acceleration Index—and findings emerging from these measures
- **Rosemarie Ziedonis** (Boston University): Insights from work conducted with Bo Zhao on an R&D loan program in Michigan. Numerous states have launched similar and large-scale programs over the past few decades, yet there are major data roadblocks in obtaining access to the basic pieces of information required to study the effects (e.g., difficulties measuring "innovative activity" of the small companies using standard patent-based measures
- **Sheryl Winston Smith** (Temple University): Measuring micro-foundations of innovation and entrepreneurship, and implications of early funding, intangibles, and peer effects. Leveraging deep, novel data on STEM-based entrepreneurship from earliest stages to construct novel measures of founder, startup, and early hiring characteristics. Findings emerging from research on accelerators and professional angel groups pointing to early influences on growth, hiring, funding, and exit pathways for startups. Implications of geographic and regional measures
- **Thomas Guevara** (Economic Development Administration): Cluster mapping of industry and business environments for use in making strategic

investments, recruiting new companies, and laying the groundwork for new industries. **[15 minutes each]**

Open Discussion [15 minutes]

10:45 ***Break*** *(refreshments available outside conference room 120)*

11:00 Innovation Measurement Agendas of the Future

Going forward, new measures of science and technology describing inputs and outcomes associated with innovative activity will be needed, and multiple data modes will be called upon. Commercial data, some recoverable from Web scraping and other computer science methods, may shape future measurement in a range of areas—e.g., new product introduction, quality change, prices and productivity, product diffusion—where large datasets provide advantage in terms of granularity, timeliness, geographic specificity, etc. To what extent can the digital revolution transform metrics in the area of innovation measurement? What are the roles of specialized surveys in innovative data collection? Administrative records (e.g., linking patent information to other data sources)? **[15 minutes each]**

- **Scott Stern** (***session lead***): Innovation that is cumulative vs that which is one off; structural characteristics of innovation. Is it possible to develop statistics that provide clues about application and diffusion of innovation?
- **Jeff Oldham** (Google): Tools beyond surveys instruments that can be used to measure innovation and productivity. Interesting work on patent data for target users
- **Ron Jarmin** (U.S. Census Bureau): Goals/actions/policies that the statistical agency data programs might strive for going forward to measure business dynamism and innovation in high tech industries. Partnerships and collaborations between the agencies
- **Sallie Keller** (Social and Decision Analytics Laboratory): Evaluating nonsurvey data for statistical use; advancing statistical and quantitative data-gathering techniques to enhance research interests

Open Discussion [15 minutes]

12:30 pm ***Lunch*** *(available outside conference room 120)*

Synthesis and Directions: Shaping Innovation Data and Indicators for the Future

Scott and **Bronwyn** will facilitate discussion summarizing guidance to NCSES as it prepares for the OECD Blue Sky III Conference. Insights aimed at broadening and modernizing innovation indicators—through advancement of measurement frameworks, identification of data gaps, and exploitation of new data sources—will be distilled. The goal is to emerge from the meeting envisioning priorities for NCSES data collection programs as the agency seeks to make its statistics relevant to data users and producers across measurement, academic, and practitioner communities spanning both the public and private sectors.

2:00 pm *Planned Adjournment*

Appendix B

Workshop Participants

STEERING COMMITTEE

Scott Stern (Chair), *David Sarnoff Professor of Management of Technology, MIT Sloan School of Management*

Ashish Arora, *Rex D. Adams Professor, The Fuqua School of Business, Duke University*

Maryann Feldman, *S.K. Heninger Distinguished Professor, Department of Public Policy, University of North Carolina at Chapel Hill*

PRESENTERS

Rajshree Agarwal, *Rudolph P. Lamone Chair and Professor of Entrepreneurship, Robert H. Smith School of Business, University of Maryland*

Wesley Cohen, *Frederick C. Joerg Professor and Professor of Economics, The Fuqua School of Business, Duke University*

Charles Edquist, *Rausing Professor in Innovation Studies at CIRCLE, Lund University, Lund, Sweden*

Cathy Fazio, *Managing Director, Laboratory for Innovation Science and Policy, MIT Innovation Initiative*

Kaye Husbands Fealing, *Chair and Professor, School of Public Policy, Georgia Institute of Technology*

Lee Fleming, *Faculty Director, Coleman Fung Chair in Engineering Leadership, Haas School of Business, University of California, Berkeley*

Fernando Galindo-Rueda, *Senior Economist, Economic Analysis and Statistics Division, OECD Directorate for Science, Technology, and Industry, Paris, France*
Alfonso Gambardella, *Professor of Management, Department of Management and Technology, Bocconi University, Milan, Italy*
Fred Gault, *Professorial Fellow, UNU Merit, Maastricht, The Netherlands*
Thomas Guevara, *Deputy Assistant Secretary, Office of Regional Affairs, Economic Development Administration, U.S. Department of Commerce*
Bronwyn Hall, *Professor of Economics (emerita), University of California, Berkeley*
Ron Jarmin, *Assistant Director for Research and Methodology, Office of the Associate Director of Research and Methodology, U.S. Census Bureau*
Sallie Keller, *Director and Professor of Statistics, Social and Decision Analytics Lab, Biocomplexity Institute of Virginia Tech*
Ben Martin, *Professor of Science and Technology Policy Studies, Science Policy Research Unit, Business and Management, University of Sussex, UK*
Kristin McCue, *Principal Economist, Center for Economic Studies, U.S. Census Bureau*
Javier Miranda, *Principal Economist, Center for Economic Studies, U.S. Census Bureau*
Jeff Oldham, *Software Engineer, Google, Mountain View, California*
Jason Owen-Smith, *Executive Director, Institute for Research on Innovation and Science, and Research Professor, Survey Research Center, University of Michigan*
Daniel Sarewitz, *Senior Sustainability Scientist, Julie Ann Wrigley Global Institute of Sustainability, and Professor of Science and Society, School of Life Sciences, College of Liberal Arts and Sciences, Arizona State University*
Robert Seamans, *Associate Professor of Management and Organizations, Leonard N. Stern School of Business, New York University*
Dan Sichel, *Professor of Economics, Wellesley College*
Paula Stephan, *Professor of Economics, Georgia State University*
Bruce Tether, *Professor of Innovation Management and Strategy, Department of Innovation, Management, and Policy, Manchester Business School, UK*
Eric von Hippel, *Professor of Technological Innovation, MIT Sloan School of Management*
Sheryl Winston Smith, *Assistant Professor of Strategic Management and Entrepreneurship, Fox School of Business, Temple University*
Rosemarie Ziedonis, *Associate Professor, Strategy and Innovation, Questrom School of Business, Boston University*

SPONSOR

Mark Boroush, *Senior Analyst, National Center for Science and Engineering Statistics, National Science Foundation*
Jaqui Falkenheim, *Senior Analyst, National Center for Science and Engineering Statistics, National Science Foundation*
John Gawalt, *Division Director, National Center for Science and Engineering Statistics, National Science Foundation*
Derek Hill, *Sciences Resources Analyst, National Center for Science and Engineering Statistics, National Science Foundation*
John Jankowski, *R&D Statistics Program, National Center for Science and Engineering Statistics, National Science Foundation*
Audrey Kindlon, *National Center for Science and Engineering Statistics, National Science Foundation*
Joydip Kundu, *Acting Deputy Division Director, National Center for Science and Engineering Statistics, National Science Foundation*
Kelly Phou, *Survey Statistician, National Center for Science and Engineering Statistics, National Science Foundation*
Carol Robbins, *Senior Analyst, National Center for Science and Engineering Statistics, National Science Foundation*

GUESTS

Jeff Alexander, *SRI International, Arlington, Virginia*
Gary Anderson, *Economist, Office of the Director, National Institute of Standards and Technology*
Xavier Cirera, *Senior Economist, Innovation and Entrepreneurship, The World Bank*
Genevieve Croft, *Senior Associate for Research Policy, Association of Public and Land-grant Universities, Washington, DC*
Louise Earl, *Section Chief, Science and Innovation, Statistics Canada*
Lucia Foster, *Chief of Center for Economic Studies, Chief Economist, U.S. Census Bureau*
Christina Freyman, *Science and Technology Policy Analyst, Center for Science, Technology, and Economic Development, SRI International, Arlington, Virginia*
Nathan Goldschlag, *Economist, Center for Economic Studies, U.S. Census Bureau*
Rasmus Hartmann, *Postdoctoral Fellow, Behavioral and Policy Sciences, MIT*
Cassandra Ingram, *Economist, Office of the Chief Economist, Economics and Statistics Administration, U.S. Department of Commerce*
Kei Koizumi, *Assistant Director for Federal R&D, Office of Science and Technology Policy, Washington, DC*

Vicki Lancaster, *Social and Decision Analytics Lab, Biocomplexity Institute of Virginia Tech*
Wendy Li, *Office of the Director, Bureau of Economic Analysis, U.S. Department of Commerce*
Vladimir Lopez-Bassols, *Formerly of OECD*
Haig McCarrell, *Director, Investment, Science, and Technology, Statistics Canada*
Oscar Mendez, *Program Associate, Economics, Alfred P. Sloan Foundation, New York, NY*
Silvia Muzi, *The World Bank*
Eric Smith, *Senior Policy Advisor, conomic Development Administration, U.S. Department of Commerce, Washington, DC*
Stephanie Shipp, *Deputy Director and Research Professor, Social and Decision Analytics Lab, Biocomplexity Institute of Virginia Tech*
Josh Trapani, *Associate Vice President for Research and Policy Analysis, Association of American Universities, Washington, DC*
Tim Wojan, *Senior Economist, Economic Research Service, USDA, Washington, DC*
Arvids Ziedonis, *Visiting Professor, Strategy and Innovation, Questrom School of Business, Boston University*
Kathryn Ziemer, *Social and Decision Analytics Lab, Biocomplexity Institute of Virginia Tech*

STAFF

Brian Harris-Kojetin, *Deputy Director, Committee on National Statistics, The National Academies of Sciences, Engineering, and Medicine, Washington, DC*
Chris Mackie, *Study Director, Committee on National Statistics, The National Academies of Sciences, Engineering, and Medicine, Washington, DC*
Anthony Mann, *Program Coordinator, Committee on National Statistics, The National Academies of Sciences, Engineering, and Medicine, Washington, DC*

COMMITTEE ON NATIONAL STATISTICS

The Committee on National Statistics was established in 1972 at the National Academies of Sciences, Engineering, and Medicine to improve the statistical methods and information on which public policy decisions are based. The committee carries out studies, workshops, and other activities to foster better measures and fuller understanding of the economy, the environment, public health, crime, education, immigration, poverty, welfare, and other public policy issues. It also evaluates ongoing statistical programs and tracks the statistical policy and coordinating activities of the federal government, serving a unique role at the intersection of statistics and public policy. The committee's work is supported by a consortium of federal agencies through a National Science Foundation grant.